Fundamentals of Connected and Automated Vehicles

Fundamentals of Connected and Automated Vehicles

BY JEFFREY WISHART
AND
YAN CHEN, STEVEN COMO, NARAYANAN KIDAMBI, DUO LU AND YEZHOU YANG

Victoria Wishart - Editorial Consultant

Warrendale, Pennsylvania, USA

400 Commonwealth Drive
Warrendale, PA 15096-0001 USA
E-mail: CustomerService@sae.org
Phone: 877-606-7323 (inside USA and Canada)
724-776-4970 (outside USA)
Fax: 724-776-0790

Library of Congress Catalog Number 2021952998
http://dx.doi.org/10.4271/9780768099829

ISBN-Print 978-0-7680-9980-5
ISBN-PDF 978-0-7680-9982-9
ISBN-epub 978-0-7680-9984-3

To purchase bulk quantities, please contact: SAE Customer Service

E-mail: CustomerService@sae.org
Phone: 877-606-7323 (inside USA and Canada)
724-776-4970 (outside USA)
Fax: 724-776-0790

Visit the SAE International Bookstore at books.sae.org

Chief Growth Officer
Frank Menchaca

Publisher
Sherry Dickinson Nigam

Development Editor
Publishers Solutions, LCC
Albany, NY

Director of Content Management
Kelli Zilko

Production and Manufacturing Associate
Erin Mendicino

Contents

Foreword

Human imagination allows us to contemplate the future and anticipate technologies that solve life's challenges. Imagination motivates research and development that demonstrates aspects of what is possible in prototypes well before they are economically feasible. These prototypes establish goals that are eventually realized when there is a maturation and convergence of the required underlying technologies.

Communications and mobility are fundamental to human existence and, therefore, are often the focus of our technological aspirations. Milestone achievements such as the telephone, radio, automobile, and airplane have had a huge impact on human existence. Their realization has, in turn, inspired new science fiction about pervasive wireless communications and automation that promises to further extend human reach, establish new levels of comfort and efficiency, and reduce unwanted side effects of existing technologies such as the accident rates we tolerate with modern transportation. A great example of how science fiction has prophesied technology development in the communications space, and where the required underlying technologies have matured over time to allow this vision to be economically realized, is Dick Tracy's two-way radio wristwatch (circa 1965), which is now a reality made possible by advances in semiconductors, computer and display design, battery technology, and communications algorithms.

Advances in computing, data processing, and artificial intelligence (deep learning in particular) are driving the development of new levels of automation that will impact all aspects of our lives. Profit motive will dictate which forms of automation will be realized first, and the importance of mobility is aligning significant resources behind the development of Connected and Automated vehicles (CAVs). What are CAVs and what are the underlying technologies that need to mature and converge for them to be widely deployed? "Fundamentals of Connected and Automated Vehicles" is written to answer these questions, providing deep insight into CAV design and the underlying technologies involved, educating the reader with the information required to make informed predictions of how and when CAVs will impact their lives.

All of the authors of "Fundamentals of Connected and Automated Vehicles" are researchers involved in the Institute of Automated Mobility (IAM), which was established in 2018 to "Provide the technical guidance and coordination required to ensure the prudent implementation of safe, efficient automated mobility across Arizona." The IAM has deep connections across the CAV community where it is recognized as a leader in the development of CAV safety-assessment technology, in particular for its pioneering research into operational safety metrics. The successful launch of CAVs will involve the application of automation advances to all corners of our roadway transportation systems, both in automobiles and across the entire transportation infrastructure. The participation of the authors in the IAM affords them a unique perspective on how automation technology will be broadly implemented, allowing them to communicate to the reader the full extent of how the technologies involved will be integrated into, and impact, their lives. I am honored to be the Technical Director of the IAM where I am humbled by the expertise of the authors and the contributions they are making to the advancement of the launch of the age of CAVs.

Greg Leeming
Institute of Automated Mobility (IAM) - Technical Director

Preface

"The automotive industry is in flux." The opening line of this book is, if anything, an understatement. Changes are impacting every aspect of this industry, from vehicle design to ownership models to safety, and connectivity and automation are two of the biggest change agents. The automotive industry is arguably transforming more rapidly in 2021 than at any other point since Carl Benz applied for his patent on a "vehicle powered by a gas engine" in 1886.

The purpose of this book is to provide the reader with an understanding of the scope and breadth of change in the automotive industry being wrought by connectivity and automation. This understanding can be achieved through an exploration of the Fundamentals of Connected and Automated Vehicles (CAVs). These fundamentals include the historical development and context (Chapter 1) and the technologies involved (Chapters 2-7) and also the verification and validation (Chapter 8) steps required to ensure that CAVs are safely deployed on public roads is also described. Finally, a perspective on the outlook for CAVs, particularly in various use cases, is provided (Chapter 9) to give the reader a sense of the timing of CAVs.

Each chapter of this book is worthy of its own, devoted book. However, each chapter contains sufficient technical depth to the reader to allow for a fundamental understanding of the topic of each chapter. The intended audience is the Engineering student enrolled in a class on CAVs, the automotive industry member who wants to gain an understanding of the major industry trends, or the layperson who is interested in delving more deeply into the CAV topic than is generally covered in media. The authors hope that they have succeeded in providing a comprehensive yet accessible guide to CAVs, one of the most transformative technological developments in the history of humankind's ingenuity.

VOL
MAP
RADIO
MEDIA
NAV
FRONT
REAR
PASSENGER AIR BAG
CLIMATE
MODE
DUAL
A/C
CleanAir
TEMP
AUTO
TEMP

1

Introduction and History of Connected and Automated Vehicles

The automotive industry is in flux. There are changes occurring that are transforming the industry in more ways and to a greater degree than anything that has occurred since Henry Ford introduced mass production of the automobile with the Model T in 1913. To date, since the internal combustion engine (ICE) became the dominant powertrain focus (beating electricity and steam engines) due, in no small part, to the dominance of the Model T and the introduction of the electric starter motor, the basic automotive design paradigm has not changed significantly. Vehicles have become more efficient and less polluting, and there are many more features and amenities packed into the interiors, including safety systems and devices that have greatly reduced the number of roadway deaths per mile traveled. The computing power on board vehicles now allows for much greater control and information being accessible about the status of the various components. However, the ICE powertrain remains dominant and the basic premise of a single, human driver controlling the vehicle speed and trajectory using a steering wheel and acceleration, brake, and (for manual transmissions) clutch pedals is unchanged. Communication between drivers is minimal, with signal lights being the main means for indicating driver intention. Finally, despite efforts by various regulators to increase the average number of occupants in a vehicle through high-occupancy vehicle (HOV) lanes and other incentive programs, single-occupancy vehicles continue to be an issue impacting traffic congestion and the environment; an estimated 76.3% of U.S. commuters drive alone without any passengers in their vehicles, vastly reducing the efficiency of traffic flow (Puentes, 2017).

This overall paradigm stasis is rapidly changing, however. Today's automotive industry is grappling with challenges associated with the mass adoption of vehicles, including rapid adoption in the developing world. There is a need to further increase vehicular safety, reduce the environmental impact of vehicle use, and improve traffic congestion. Four dominant industry trends, collectively known as CASE, shown below in the Venn diagram of Figure 1.1, seek to address several of these challenges:

- Electrification
- Connectivity
- Automation
- Sharing

FIGURE 1.1 Current automotive industry trends.

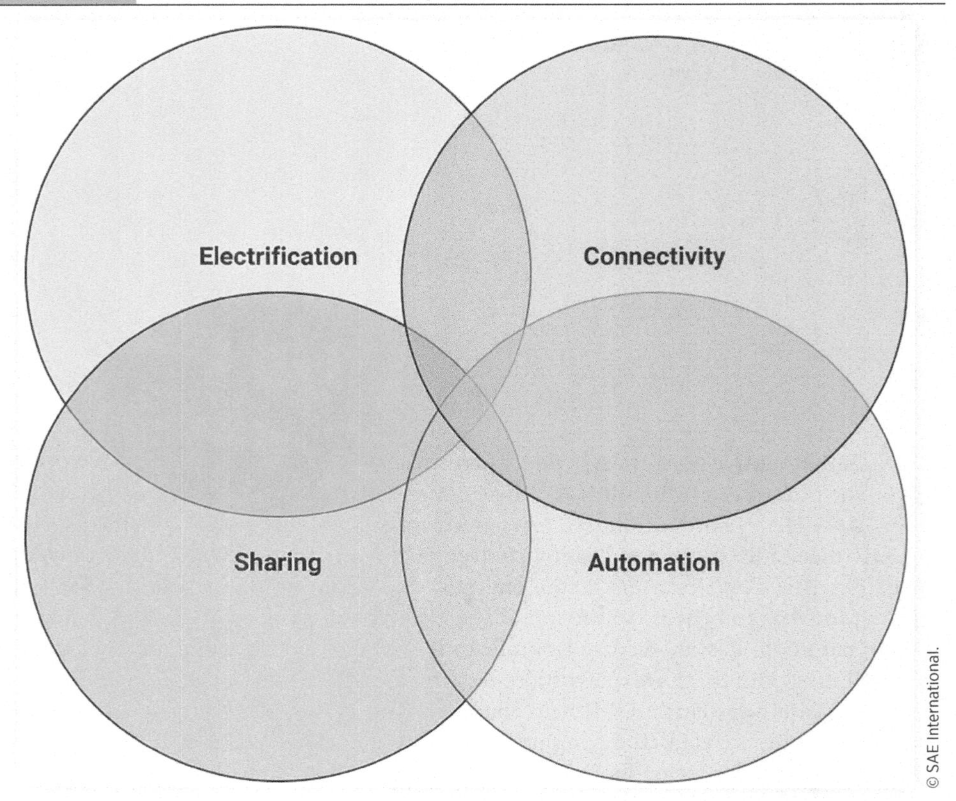

Electrification refers to moving away from the ICE as the sole source of motive power toward powertrains that incorporate electric motive power (via electric motors [EMs]), and potentially complete electric motive power in the form of electric vehicles (EVs). Electrification is the most mature of the four trends. Given the dominance of ICEs in today's marketplace, it may be surprising to some that EVs have been built sporadically in the decades since the early twentieth century when EVs competed with ICE vehicles and steam-powered vehicles for early automobile powertrain dominance. However, the commencement of sustained, significant R&D work in vehicle electrification is only two decades old. EVs accounted for 1% of new U.S. vehicle sales in 2018, though the percentage is 5% in California and is higher in some other countries, most notably Norway, where the percentage was 54% in 2020. Increasing electrification is seen as inevitable for several reasons. EVs are substantially more efficient at converting energy to vehicle motion than ICE-based vehicles

(~83% vs. 22% [Ramachandran & Stimming, 2015]); have fewer moving, mechanical parts (making them less costly to maintain); and offer the ability to drastically reduce emissions at the tailpipe and from a so-called well-to-wheels basis if the electricity is produced with cleaner fuel sources. Fortunately, the latter is another prevailing trend, as energy grids worldwide increase their proportion of renewable sources and governmental policies lean toward increased environmental responsibility.

Connectivity is the next most mature trend, although there is some controversy within the automotive industry as to what exactly the term means. For the purposes of this book, the term will refer to any vehicle that can send or receive communications from off-board sources. Examples include global positioning system (GPS) signals, cellular communications, and vehicle-to-everything (V2X). These terms of art will be explored in later chapters. GPS-equipped vehicles were introduced by Oldsmobile circa 1994 (Mateja, 1995), and certain contemporary vehicles can receive Signal Phase and Timing (SPaT) information from traffic signals that provide information on the traffic light sequence and timing. Connectivity allows for much richer information to be provided to a vehicle driver (or automated driving system [ADS]), such that more informed decisions can be made. For example, if a driver is alerted to the presence of a disabled vehicle downstream on a curve that is occluded by foliage, the driver can alter his/her speed to anticipate the lane being blocked and, thus, avoid an accident. Better decisions can be made in a myriad of situations, and Connectivity is a key link for vehicles to an era where data, and data science, play an increasingly important role in modern life.

Sharing is the second-least mature trend yet is familiar to most people today, though the concept is currently being expanded using ICE vehicles generally, without Electrification, Connectivity, or Automation. Sharing includes ride hailing (exemplified by companies like Lyft and Uber) and car sharing (provided by companies like Zipcar and AutoShare). Sharing also includes various modes of transportation, known as multi-modal transportation, from scooters to bicycles to cars to buses and trains; trips can often use multiple modes to arrive at a destination. Sharing allows for lower levels of vehicle ownership, which could be quite disruptive to the automotive industry. Sharing can also provide enhanced mobility for individuals who cannot drive or prefer not to drive, including disabled persons, seniors, children, and those who would like expanded mobility options.

Last in maturity, but certainly not in potential for automotive industry sector disruption, is automation. Automation in the automotive context refers to vehicles with driving automation systems.[1] Vehicles equipped with ADS can have varying levels of driving automation features. These range from features such as adaptive cruise control, labeled as Driver Assistance by the SAE J3016 document (Society of Automotive Engineers, 2018) (discussed further in Chapter 2), to Full Automation, where the vehicle's steering wheel and pedals are optional and no human supervision is required. Allowing the ADS greater control over the driving task will allow human occupants to spend time in the vehicle engaged in tasks other than driving, including sleeping, working, watching a video, or even exercising. Automation also has the strong likelihood of significantly reducing the frequency of accidents and, like Sharing, increasing mobility for those who cannot drive themselves. Automation is arguably the biggest change in the vehicle industry since the invention of the vehicle itself.

The relative importance of each trend shown in Figure 1.1 on the automotive industry as a whole is yet to be defined or understood. Further, the overlaps between the trends range

[1] According to the SAE J3016 nomenclature, a driving automation system refers to automation from Level 1 through Level 5 while ADS refers to automation from Level 3 to Level 5 only.

from obvious to nebulous. For example, wireless charging might enable more vehicles with an ADS to be EVs, in an Electrification-Automation overlap. However, the connection between Connectivity and Electrification is perhaps less clear; on the one hand, knowing that an electric vehicle supply equipment (EVSE) unit is available is crucial to the EV driver; however, this information can come from the driver's smartphone and is not necessarily required to be provided by the vehicle display. How each trend develops and grows will likely be highly dependent on the overlap with and the development of the other trends.

This book focuses on two of the four trends: Connectivity and Automation. There is obvious overlap between the two, and the synergy created through leveraging the advantages of both trends may dramatically alter the automotive landscape such that in the not-too-distant future, connected and automated vehicles (CAVs) could become dominant.[2]

CAV History and Origins[3]

The first automated vehicles were arguably developed in ancient times: sailboats have a weathervane connected to the tiller with ropes to keep the sailboat traveling on a single course even if the winds shift (Figure 1.2).

FIGURE 1.2 Sailboats were the original CAVs.

Hayden Terjeson/Shutterstock.com

[2] It should be noted that not all vehicles equipped with connectivity technology will also be equipped with automation technology, and vice versa. The CAV term will be used in both situations interchangeably.

[3] Some content in this section is sourced from Weber (2014).

The history of motive automation involving human-occupied vehicles was then revisited with the invention of the autopilot system for airplanes that allowed for simultaneous flying and navigation. The Sperry Gyroscope Autopilot developed in the 1930s is shown in Figure 1.3. This device allowed Wiley Post to fly solo around the world in 1933.

FIGURE 1.3 The Sperry Gyroscope Autopilot.

Motive automation was also advanced (in a recurring theme) for military applications: Torpedoes, such as the version dropped by a low-flying airplane depicted in Figure 1.4, were first developed in the 1860s with simple guidance systems that maintained course and depth. By World War II, sonar capabilities were added so that specific targets could be attacked.

FIGURE 1.4 Torpedo dropped by a low-flying airplane.

Military incentives kept up the automation development, and the German V2 rocket, shown in Figure 1.5, was the first human-made equipment to travel in outer space, with its course directed by a series of gyroscopes.

FIGURE 1.5 German V2 rocket.

Reprinted from Imperial War Museum.

The idea of automated passenger vehicles first appeared in comic books and science fiction novels, starting circa 1935. Norman Bel Geddes of General Motors (GM) was the first original equipment manufacturer (OEM) employee to publicly describe a vision of automated vehicles in his Futurama ride at New York's World Fair in 1939. The vision included speed and collision controls similar to railroads and "trenches" to keep vehicles in their lanes. Of course, without computers, this vision was fantastical. But these early depictions elucidated benefits of CAVs that continue to be attractive today, such as the reduced need for parking spaces in downtown areas (since CAVs can drop the passengers off and park outside of the downtown core), reduced accidents, higher efficiency/road density, and more family time. As stated in the fine print in Figure 1.6, commutes could be productive resulting in shorter workdays, and road trips can involve interactions between the driver and passengers (i.e., parent and children, respectively) when the seating configuration allows it.

FIGURE 1.6 Early depiction of a CAV and the promised benefits.

Radio Corporation of America (RCA) and GM conducted tests such as the one shown in Figure 1.7 in the 1950s with vehicles that had radio receivers, magnets, and steel cables embedded in the road for steering and speed control. The GM vehicles had "pick-up" coils at the front that detected current flowing in the embedded cables, with different current values directing the vehicle to turn left or right. The control tower transmitted radio signals for speed control.

Using the advances in computing, the earliest CAV attempts were aimed at "cockroach-like" motion:

- Sensing
- Processing
- Reacting

Sensing and reacting were possible with contemporaneous technology. But the machine intelligence of the processing step needed to be developed. Stanford University was among the pioneers in the 1960s and 1970s for robotics and machine vision, developing the Stanford Artificial Intelligence Cart shown in Figure 1.7.

FIGURE 1.7 Stanford Artificial Intelligence Cart.

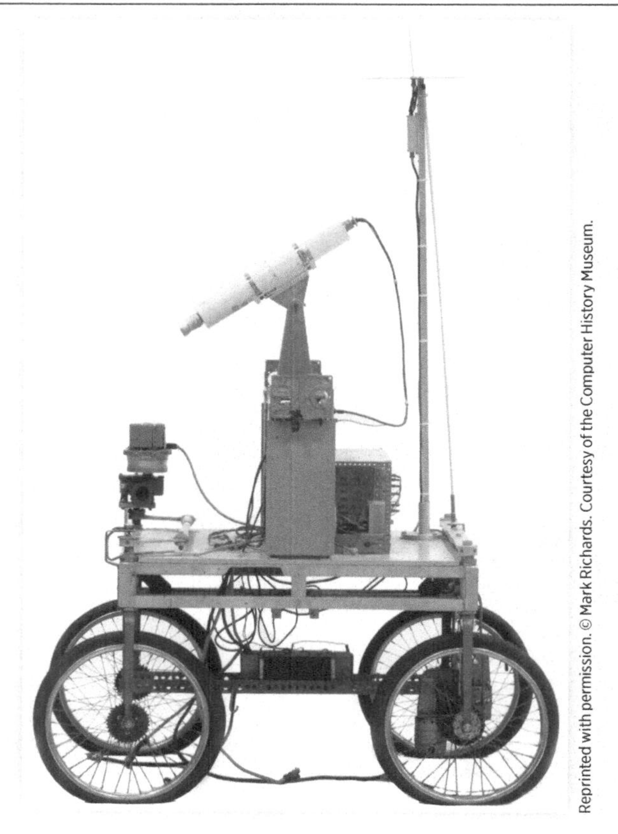

Reprinted with permission. © Mark Richards. Courtesy of the Computer History Museum.

By 1980, Ernst Dickmanns, a professor at the Bundeswehr University in Munich, had developed a Mercedes van that was able to travel hundreds of highway miles with a primitive version of automated driving. By 1993, Dickmanns' VaMP Mercedes sedan, part of the Eureka PROMETHEUS project, could travel in traffic at speeds up to 100 mph. Figure 1.8 is an image of the 1997 version of the Dickmanns VaMoRs van, with the third generation of vision systems developed.

FIGURE 1.8 Ernst Dickmanns' automated van.

Reprinted with permission. © Dynamic Machine Vision.

Military applications continued to push CAV development. The Predator drone, shown in Figure 1.9, has been used for surveillance since at least 1995 and for targeted strikes since at least 2001, though the precise dates of the first deployment are not publicly known.

FIGURE 1.9 Predator drone.

Oleg Yarko [site: Yarko.Tv]/Shutterstock.com.

Autonomous Underwater Vehicles (AUVs), depicted in Figure 1.10, were also under development for both military applications and scientific endeavors like mapping the ocean floor. AUVs can also be used in marine search applications. The Kongsberg HUGIN 4500 units aboard the MF Fugro Supporter and MV Havila Harmony vessels are self-propelled and equipped with cameras, and were part of the search for remains from the Malaysia Airlines Flight 370 at depths up to 20,000 ft (Joint Agency Coordination Centre, 2015).

FIGURE 1.10 Autonomous Underwater Vehicle.

Reprinted with permission. © Bluefin.

In 1995, the Carnegie Mellon University (CMU) NavLab 5 team's CAV, shown in Figure 1.11, traveled from Pittsburgh to San Diego. Averaging above 60 mph, 98% of the journey was completed by the automation system (approximately 2,740 CAV miles).

FIGURE 1.11 Carnegie Mellon NavLab 5.

Reprinted with permission. © Carnegie Mellon University.

In 1991, the U.S. Congress authorized $650M to support the development of automated highway driving. By 1997, CAVs were being tested on 7.6 miles of the HOV lane of Interstate 15 near San Diego, as shown in Figure 1.12. This trial, and others like it, received some public attention but did not lead to a sustained effort in the automotive industry to develop CAVs.

FIGURE 1.12 CAV testing near San Diego in 1997.

CAV development truly became widespread with the introduction of the Defense Advanced Research Projects Agency (DARPA) Grand Challenge for CAVs. The DARPA Grand Challenge was initiated in 2004 as the first long-distance (150 miles) race in the world for these vehicles. The grand prize was $1M. In 2004, no vehicles finished the course; the furthest any team traveled was 7.3 miles before getting "hung up" on some rocks. In 2005, a total of 45 vehicles entered the competition. A total of 22 out of 23 finalist vehicles surpassed the 7.3-mile mark, and five vehicles successfully completed the course. The winner was Stanley, shown in Figure 1.13, developed by Stanford University and completed the course in a time of 6 h 54 min.

FIGURE 1.13 Stanford's Stanley CAV, winner of the 2005 DARPA Challenge.

The DARPA Grand Challenge was modified for the 2007 race and became the "Urban Challenge." The course was 60 miles of a simulated urban environment and had to be completed in under 6 h. All traffic laws had to be obeyed. The Grand Prize was $2M and some teams received up to $1M in funding in order to help with development. The Urban Challenge forced the CAVs to make more frequent real-time decisions with respect to other vehicles. The winning vehicle, named "Boss" and shown in Figure 1.14, was developed by Tartan Racing, a team led by CMU.

FIGURE 1.14 CMU and GM joint venture, Tartan Racing, winner of the 2007 DARPA Urban Challenge.

Reprinted from Public Domain.

Universities continued to lead CAV development around this time. The University of Parma developed the VisLab CAV, shown in Figure 1.15. In 2010, it undertook what could be the most geographically diverse automated driving route to date: 10,000 miles through 9 countries in 100 days from Parma, Italy, to Shanghai, China. This vehicle is reputed to be the first CAV to receive a traffic ticket. The police officer reportedly did not know whose name to put on the ticket (Vanderbilt, 2012).

FIGURE 1.15 University of Parma's VisLab CAV.

Reprinted with permission. © VisLab.

CAVs Today

From relatively modest beginnings, CAVs are commercially available today for specific applications, such as forklifts and low-speed shuttles, shown in Figures 1.16 and 1.17, respectively.

FIGURE 1.16 Automated forklift.

Chesky/Shutterstock.com.

FIGURE 1.17 Local Motors' Olli electric, automated shuttle.

Reprinted with permission. © Local Motors.

Induct Technology produced what may have been the first commercially available on-road CAV, the Navya, shown in Figure 1.18. It had a top speed of 12 mph, a cost of 250k USD, and a very limited set of conditions under which it could operate (known as an Operational Design Domain (ODD), which will be further discussed later in the chapter).

FIGURE 1.18 Induct Technology's Navya CAV, the first commercially available CAV.

Reprinted with permission. © Navya.

London Heathrow Airport has demonstrated the "robo-taxi" from Ultra Global PRT, shown in Figure 1.19.

FIGURE 1.19 London Heathrow Airport's robo-taxis from Ultra Global PRT.

For the light-duty vehicle market, the CAV industry includes both established automotive industry companies, established companies in other technology areas, and start-ups:

- OEMs (e.g., General Motors, Ford, Volkswagen)
- Tier 1 suppliers (e.g., Denso, Bosch, ZF)
- Silicon Valley (e.g., Waymo, Zoox, Aptiv)

The field of competitors vying for commercial introduction of CAVs is broad and turbulent. Many companies are acquiring other companies or joining together to form consortia and partnerships to help speed development times and share the large costs:

- Bayerische Motoren Werke (BMW)-Intel-Delphi-Mobileye
- GM-Cruise Automation (shown in Figure 1.20)
- Uber-Volvo-Microsoft-Autoliv
- Waymo-Fiat Chrysler Automobiles (FCA) (shown in Figure 1.21)
- Audi-Nvidia

FIGURE 1.20 GM-Cruise prototype CAV.

FIGURE 1.21 Waymo-FCA test CAV.

Reprinted with permission. Photo by Daniel Lu.

Determining the leading companies and consortia is an inexact science, but Guidehouse (formerly Navigant, a consulting company that releases reports on market analysis) has provided CAV industry rankings since 2017. The rankings criteria are:

- Vision
- Go-to market strategy
- Partners
- Production strategy
- Technology
- Sales, marketing, and distribution
- Product capability
- Product quality and reliability
- Product portfolio
- Staying power

Guidehouse's CAV leaderboards in January 2017-February 2021 are shown in Figures 1.22 to 1.26. Note that in the February 2021 leaderboard, Guidehouse modified the analyzed companies to only include CAV developers, and not companies that will only be deploying CAVs without also developing the vehicles.

FIGURE 1.22 Guidehouse's CAV leaderboard: January 2017.

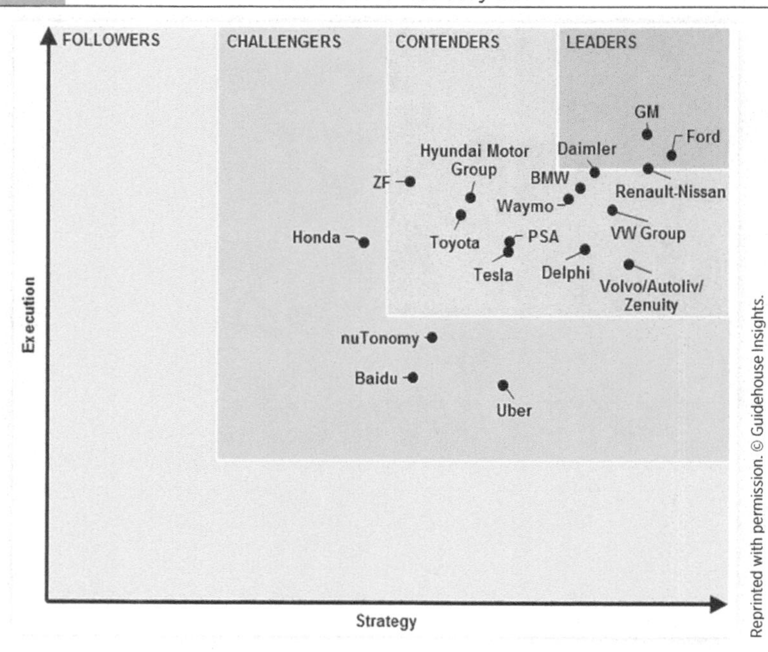

FIGURE 1.23 Guidehouse's CAV leaderboard: January 2018.

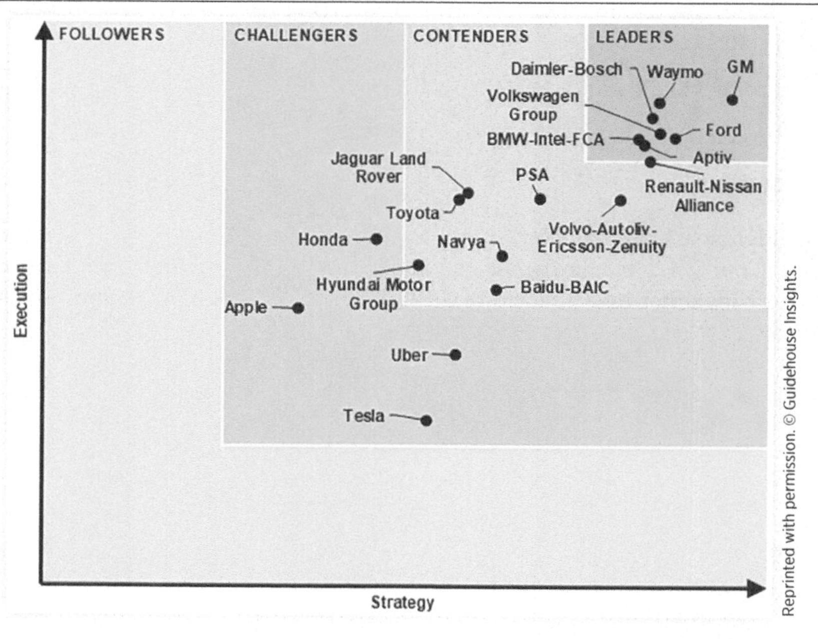

FIGURE 1.24 Guidehouse's CAV leaderboard: January 2019.

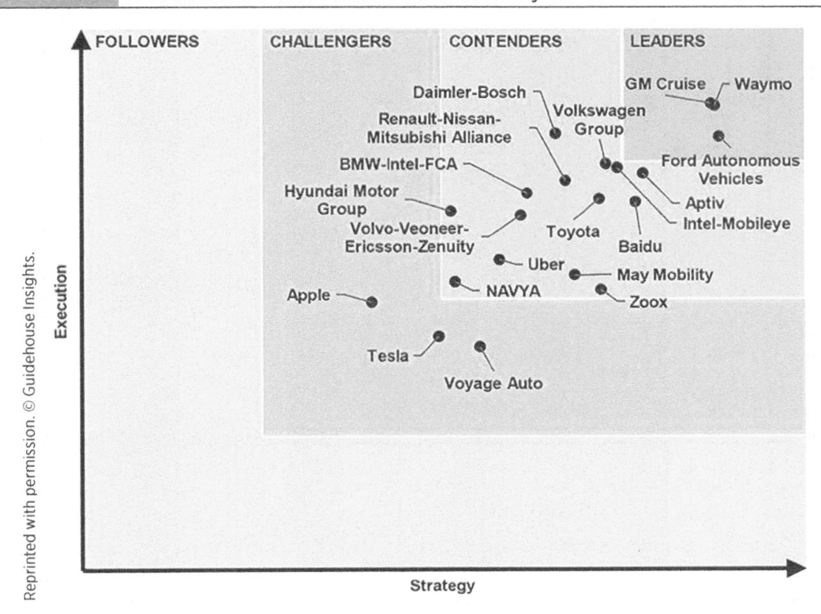

FIGURE 1.25 Guidehouse's CAV leaderboard: January 2020.

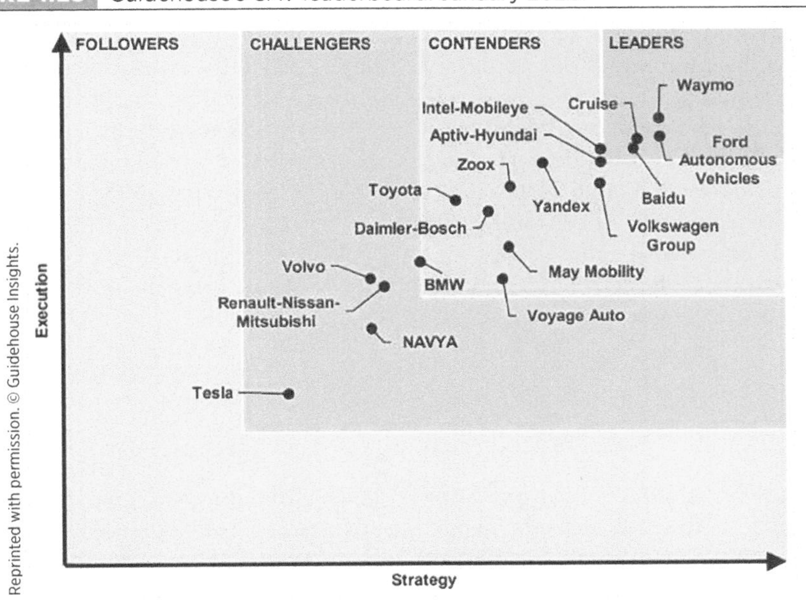

FIGURE 1.26 Guidehouse's CAV leaderboard: April 2021.

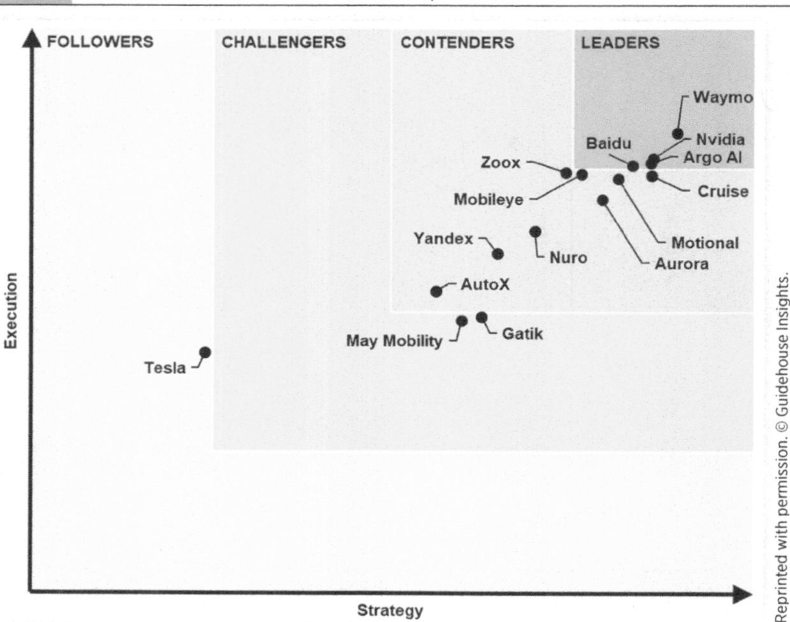

Reprinted with permission. © Guidehouse Insights.

From these rankings, it appears that Waymo has, at least in Guidehouse's estimation, a durable, but small, lead over its rivals. Other current leaders appear to be Nvidia, Argo AI (a division of Ford), and Baidu (of Chinese internet search fame). Cruise (owned by GM), Intel-Mobileye, Zoox, and the Hyundai-Aptiv joint venture Motional are on the cusp of being considered industry leaders. Some companies are present in all five (2017–2021) years considered, while others appear or disappear in different years. Of particular note is that Guidehouse has consistently ranked Tesla low in both strategy and execution, even though Tesla has a reputation, at least in the public media domain, as being in a dominant position because its vehicles are equipped with sensors that allow the company to collect a vast amount of data to train its artificial intelligence (AI) algorithms. The strategies adopted by some of these groups will be discussed in detail in future chapters.

Current Status

There is considerable excitement and interest in CAVs from industry, government, and the general public. However, there are many issues to be addressed before CAVs are deployed commercially, let alone dominate sales and in number on public roads. The anticipation that CAVs would be ready by the early 2020s has been tempered due to a greater appreciation that some obstacles are more difficult to overcome than previously understood. The predicted timeframe of widespread CAV deployment has stretched, and there are differing opinions on whether fully automated vehicles will be commercialized this decade or several decades from now.

In order to understand the current status of CAVs, it is helpful to note that there are three main systems in a CAV:

1. Perception system: Required to detect objects in the environment, classify, locate, and determine the speed and direction of each safety-relevant object. The ADS must also locate itself on its detailed, three-dimensional (3D) map of roads within its ODD (defined below).
2. Path planning system: Once the surroundings are "perceived," a plan of action must be developed based on the vehicle mission and the characteristics of the identified objects.
3. Actuation system: With a plan of action completed, the vehicle controls must be actuated in order to complete the plan of action.

The perception system is seen as the ADS sub-system requiring the most development. The perception of the CAV must improve to reduce the number of false negatives (not detecting an object) and false positives (ghost objects). Object classification must improve as well. The system must also include redundancy to account for sensor error and failure. The perception of "known" or "average" scenarios may be adequate, but the system must also be capable of handling more complex and difficult scenarios:

- Inclement weather
- Roads with temporary restrictions/temporary obstacles
- Parking lots
- Heavy pedestrian/cyclist traffic
- Non-mapped areas

Even if perfect perception of an environment was possible (this level is likely to only be asymptotically approached), the path planning system must be able to navigate scenarios that have not been "seen" in testing or training. The technology of CAVs is thus not sufficiently developed for public roads—the industry as a whole is understanding that the first 80-90% of capability has been achieved, but the last 10-20% will require more effort and development than previously thought. Sensor technologies paths are also uncertain, for example, LIght Detection and Ranging (LIDAR) sensors are seen by some CAV developers as crucial components and by others as superfluous.

The path that this technology development will take is also unclear. There are some CAV developers that are starting with simpler automation capabilities or expanding the scope of currently available driver-assistance features, incrementally increasing the capability (commercialized throughout) until a fully automated vehicle is achieved. Others are targeting fully automated vehicles from the start. The former has the advantage of gradual adoption and increasing public familiarity with CAV technology. The latter would present the CAV as a fait accompli and avoid some of the perceived problems with partial automation where humans remain in the loop. The approach that ultimately proves more successful will depend on the financial support for each, consumer reaction and acceptance, and the perception of safety.

It remains to be seen which class of vehicles will see the first widespread CAV adoption. The light-duty segment receives much of the attention, driven partly by interest in the so-called "robo-taxi" industry that would eliminate costs associated with human drivers and could thus be more likely to justify the higher cost of CAVs over a conventional vehicle.

However, many in the CAV industry see the long-haul trucking industry as the first use case for CAVs. This is in part because the incremental cost of the automation system is less of an obstacle for commercial trucks that may already cost hundreds of thousands of dollars and in part because long-haul trucks generally traverse environments such as freeways and highways, which are less complex than those typically encountered in urban driving. The last-mile transportation and delivery industries may also claim the mantle of being the first transportation segment with widespread CAV adoption. Last-mile transportation involves lower speeds and often on prescribed routes that have lower complexity. Last-mile delivery can avoid both drivers and passengers and can involve some entities such as Nuro, which has explored driving on sidewalks at low speed to reduce interaction with road traffic.

While CAV technology is far from mature, the legislation and regulatory environment is even more nascent. One complicating factor in the United States is that, historically, the federal government has regulated the vehicle while the states have regulated the driver. Since the driver and vehicle are one and the same in CAVs, the regulatory landscape is less clear. The federal government has introduced voluntary measures, such as the guidance to CAV developers to provide a Voluntary Safety Self-Assessment (VSSA), but such measures are not mandatory and there is no requirement to provide any assurance, let alone actual data, that the CAV in question is roadworthy. States have filled the gap in part by providing a patchwork of regulatory controls. However, there are no requirements that CAV developers share their data, and no state has provided a comprehensive regulatory regime to ensure public safety. Standards Development Organizations (SDOs) such as SAE International, International Electrical and Electronics Engineers (IEEE), and International Organization for Standardization (ISO) have started to provide standards and best practices that could guide CAV development and eventually lead to regulation. While there are currently duplications and contradictions among SDOs, there are efforts underway to harmonize standards and documents. This should make the regulatory rulemaking process easier and faster, and this will help provide guardrails to the CAV industry without stifling innovation.

It is clear that while much progress has been made in CAV development, there is a long way to go before society sees widespread CAV deployment and adoption. This is true for both the technology and the regulatory environment, as described above, but there are many more stakeholders in CAVs beyond industry and government. The impact on public safety that CAVs present is massive, and the industry has not yet shown that the vehicles achieve a desired level of safety; in fact, both the desired level and the manner in which the CAV can demonstrate that the level has been met are yet to be determined (discussed further in Chapter 8—Verification and Validation). In addition to public safety, there are many aspects of society that will be impacted by CAV deployment. Some of these aspects are discussed in the following section.

Societal Impacts

The impacts on society that CAVs will have, like the overall trends in the automotive industry, are currently unclear. However, there has been much speculation about expected effects, both positive and negative. While there is uncertainty about the timing and intensity of CAV's societal impacts and any discussion is prone to prognostication, this section offers an overview of the ways in which CAVs may influence safety, the environment, and quality of life.

Positive Impacts

One of the major hoped-for impacts of CAVs is to reduce the number of deaths from traffic accidents. There are some 36,000 deaths in the United States every year in traffic accidents, and one million worldwide. The National Highway Traffic Safety Administration (NHTSA) has determined that human error is a critical factor, or last event in the crash causal chain, in more than 94% of crashes (National Highway Traffic Safety Administration, 2018). If sources of a human driver's error, which include distraction, tiredness, intoxication, and poor judgment, can be reduced, the number of crashes is also expected to be reduced. Of course, this requires that CAVs do not introduce other crash causation factors.

Fewer accidents should mean a reduction in car insurance costs since the insurance companies would have fewer pay-outs to amortize over the premiums paid by individual drivers. Reduced insurance costs would make owning a vehicle more affordable for many people who cannot currently do so if the upfront costs of CAVs can be sufficiently reduced.

Another expected positive impact of CAVs is a reduction in traffic congestion. In the United States in 2019, congested roadways cost an estimated $88B in lost productivity and the average driver lost 99 hours during the year due to traffic (INRIX, 2020). Traffic affects both quality of life and the economic health of a country, and as urbanization continues apace across the globe (going from 55% of the global population in 2018 to an expected 68% by 2050 [United Nations Department of Economic and Social Affairs, 2018]), the problem is likely to get worse. CAVs, while unlikely to be a panacea, can help alleviate traffic problems through a variety of mechanisms, such as more uniform speeds and braking, communications between traffic participants and infrastructure, and coordination at intersections.

CAVs are expected to also increase mobility for so-called "transportation-challenged" populations such as the elderly and the disabled. While the ride-hailing companies mentioned earlier can provide mobility to those who cannot drive themselves, CAVs offer the prospect that these populations can use this mode of transportation at a cost that allows them full mobility since the cost of the driver would be eliminated. More speculatively, CAVs could allow busy parents to forego driving their kids to various activities and school. This would of course depend on the level of trust parents have both in the safety of CAVs and in allowing their children to be unaccompanied; the legal ramifications of the latter would also need to be addressed.

The average U.S. commute takes 26.1 min each way (U.S. Census Bureau, 2017a), for a total commuting time of nearly an hour. If the commute took place in a CAV, the commuter would be freed from the responsibility of driving so that any number of other activities could take place, including reading, sleeping, enjoying media content, etc. However, it is also possible that the commuter could be productive from their employment perspective so that the actual time spent at the place of employment could be shortened. This could mean shorter workdays and more time to spend on activities other than work.

CAVs offer the potential for more efficient land use in urban areas. Parking spaces for personal vehicles take up an inordinate amount of real estate in cities. In U.S. city centers, some 50-60% of the total amount of space is assigned to parking spaces (Old Urbanist, 2011). In homes, garages average 264 sq. ft (for single-car) and 360 sq. ft (for double-car) (Danley's Garage World, n.d.), which is 10% and 14% of the total square footage of the average U.S. home in 2017, respectively (U.S. Census Bureau, 2017b). CAVs could drop off commuters directly at their destinations in city centers and then either continue to pick up other passengers (in a ride-hailing enterprise) or park themselves at a location outside of the city center (for an individually owned vehicle). New houses could be built without garages for housing vehicles and the space could be devoted to any number of other purposes. Apartment buildings would no longer require multiple floors of parking, which would reduce costs as well as the volumetric footprint of the building.

As mentioned above, this list is by no means exhaustive. Some potential benefits have been identified while others could materialize once CAVs have been commercialized. The extent of the benefits will vary and remain to be seen.

Negative Impacts

While there appear to be many positive aspects to CAV commercialization, the possible negative aspects must not be ignored. The most significant negative impact may be the social upheaval due to labor market disruptions and job losses. For example, there are some 3.5M truck drivers in the United States whose jobs are at risk due to CAVs (U.S. Census Bureau, 2019). The median wage for U.S. long-haul truckers in 2018 was $43,680 (U.S. Bureau of Labor Statistics, n.d.); although this wage is lower than the median U.S. wage of $52,104 in 2020 (U.S. Bureau of Labor Statistics, 2020), it is higher than the median wages for those without a high school diploma ($32,760) and those with a high school diploma but no college diploma ($41,028). Long-haul trucking and taxi services offer middle-class wages to workers without higher levels of education. More indirectly, car rental and valet jobs could be affected, and fewer roadway crashes may mean fewer vehicle repair shops. Though some of the companies associated with the development and deployment of CAVs claim that cost reductions through job elimination is not the goal of introducing automation, experience from other industries that have seen increased automation suggests that it is inevitable. In order to avoid labor unrest and increased unemployment, steps will be required to ensure a transition to CAV commercialization that does not result in massive losses of jobs.

CAVs have the potential to reduce energy consumption, especially if they encounter fewer traffic delays. They can also be programmed to engage in smoother, more efficient driving with less aggressive acceleration and braking events. However, the energy consumption of the transportation network could increase overall. The convenience of CAVs may induce new trips that may not otherwise have been taken by car. CAVs may be traveling between destinations without a driver and the energy consumption of the ADS itself has been estimated at 200 W (Baxter, et al., 2018). The U.S. Department of Energy (DOE) and several national laboratories attempted to "bookend" the impact of CAVs on the transportation sector and found a range from a reduction of 60% to an increase of 200% (Stephens, et al., 2016). The wide range shows that the ultimate impacts are unknown, but that there could be a significant increase in energy consumption. As global awareness of the need to reduce energy consumption, especially of fossil fuel-related energy, grows, the true impacts must be studied, and steps must be taken to mitigate an increase due to CAVs if at all possible.

A negative impact that is related to energy consumption is the prospect that CAVs could exacerbate urban sprawl. When the length of commute is unimportant (energy costs notwithstanding, although these are low in the United States by global standards) because the commuter can sleep, eat, or even work while commuting, said commuter may choose to live farther away from their place of employment than they would otherwise have done. In addition to increasing energy use, this often results in living in suburbs or even exurbs outside the city center, thereby reducing the density of cities. All things being equal, higher density means lower energy use and lower costs since the provision of services like roads, utilities, and public transit is easier for cities. Urban planners are attempting to reduce urban sprawl and make cities more dense and "walkable," and CAVs could counteract these efforts.

In addition to public transit potentially costing more due to urban sprawl, the use of public transit may decrease, potentially significantly. The positive feedback loop of reduced revenue and (likely) resulting reduction in service could strain public transit agencies and introduce even more mobility challenges to those who cannot afford access to private vehicles. A reduction in public transit could increase overall energy consumption in the transportation sector as well as increase traffic problems and emissions. This highlights the interaction between potential negative impacts of widespread CAV deployment; an increase in one can cause a corresponding increase in one or more of the others.

While a full examination of the ethical considerations of CAVs is beyond the scope of this book, it should be noted that these considerations are incredibly important to society and how CAVs are viewed. The well-known "trolley car problem" is applicable to the path planning decisions that CAVs will have to make, and has been explored by the Massachusetts Institute of Technology (MIT) Moral Machine, where the user is presented with various scenarios of accidents that will kill people, and the user must choose between people types and numbers (Massachusetts Institute of Technology, 2020) (See Chapter 7 for more discussion on this topic). The question must also be asked whether the CAV should protect its own occupant(s) at all costs or whether there is a larger, societal obligation. How these questions are answered by regulators and CAV developers could influence how CAVs are perceived by society at large quite dramatically.

The question of liability for CAVs remains an open question. The demarcation between when the human and automation system are in control of the vehicle (described further in the section below) and the transition between the two purviews could provide fodder for lawsuits. Even when the ADS is fully in control, the question of liability has yet to be answered. On the other hand, CAVs will be capable of storing the perception data that is almost never available in the case of human-driven vehicle crashes. Accident reconstruction experts may have a powerful tool for understanding an incident. However, it is critical that the perception data are actually stored and shared with law enforcement, the CAV owner, and the other parties in the case of an accident.

The use of CAVs by malicious actors or mischievous parties has received some attention. While there is reason to believe that the CAV developers would implement safeguards to prevent CAVs from being hacked and used as weapons, the stakes of the usual cat-and-mouse game played by cybersecurity experts and hackers are higher when actual human lives are directly at stake. The problem is exacerbated if the CAV is part of a fleet and the hack of one (or of the central system) means that the control of multiple vehicles is compromised.

From the violent to the whimsical, the question must be asked whether the CAV will bring about the end of the movie car chase scene. If an automation system that does not allow speeding is in control of the vehicle, the iconic scene from *Total Recall* notwithstanding, will the action movie be the same?

CAV Taxonomy and Definitions

Just as in any industry, it is important to have common definitions so people working throughout the industry can communicate effectively and avoid confusion associated with vague and/or misleading terminology. The taxonomy and definitions in this section will be used throughout the book.

The NHTSA, part of the U.S. Department of Transportation (U.S. DOT) originally had its own taxonomy of automation levels, released in 2013. In September 2016, the SAE On-Road Automated Driving (ORAD) Committee released SAE J3016_201609, and the NHTSA adopted the SAE taxonomy. The current version of J3016 was released in 2018, and the terminology in this document as well as others prepared by the SAE ORAD Committee will be used in this book.

The taxonomy, summarized below in Figure 1.27 and explained further in Figure 1.28, is based on the levels of a driving automation system from 1 through 5 (Level 0 being a system with no driving automation system), and are separated into two distinct categories:

- Levels 0 to 2 include automation where the driver is responsible for all or part of the dynamic driving task (DDT).
- Levels 3 to 5 include automation where the ADS is responsible for the entire DDT (while engaged).

FIGURE 1.27 SAE J3016 automation level summary.

Level	Name	Narrative definition	DDT: Sustained lateral and longitudinal vehicle motion control	DDT: OEDR	DDT fallback	ODD
Driver performs part or all of the *DDT*						
0	No Driving Automation	The performance by the *driver* of the entire *DDT*, even when enhanced by *active safety systems*.	*Driver*	*Driver*	*Driver*	n/a
1	Driver Assistance	The *sustained* and *ODD*-specific execution by a *driving automation system* of either the *lateral* or the *longitudinal vehicle motion control* subtask of the DDT (but not both simultaneously) with the expectation that the *driver* performs the remainder of the *DDT*.	*Driver* and *System*	*Driver*	*Driver*	Limited
2	Partial Driving Automation	The *sustained* and *ODD*-specific execution by a *driving automation system* of both the *lateral* and *longitudinal vehicle motion control* subtasks of the *DDT* with the expectation that the *driver* completes the *OEDR* subtask and *supervises* the *driving automation system*.	***System***	*Driver*	*Driver*	Limited
ADS ("*System*") performs the entire *DDT* (while engaged)						
3	Conditional Driving Automation	The *sustained* and *ODD*-specific performance by an *ADS* of the entire DDT with the expectation that the *DDT fallback-ready user* is *receptive* to *ADS*-issued *requests to intervene*, as well as to *DDT performance-relevant system failures* in other *vehicle* systems, and will respond appropriately.	*System*	***System***	*Fallback-ready user (becomes the driver during fallback)*	Limited
4	High Driving Automation	The *sustained* and *ODD*-specific performance by an *ADS* of the entire *DDT* and *DDT fallback* without any expectation that a *user* will respond to a *request to intervene*.	*System*	*System*	***System***	Limited
5	Full Driving Automation	The *sustained* and unconditional (i.e., not *ODD*-specific) performance by an *ADS* of the entire *DDT* and *DDT fallback* without any expectation that a *user* will respond to a *request to intervene*.	*System*	*System*	*System*	**Unlimited**

FIGURE 1.28 SAE J3016 automation level explainer.

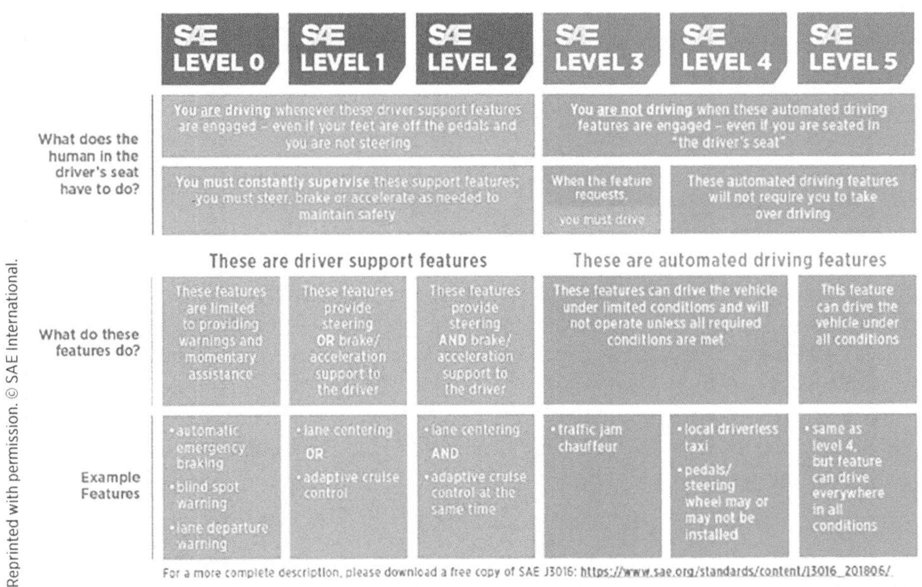

It should be noted that there is a distinction between automation and what is known as "active safety," even though some of the technologies are shared. For driving automation system features, all or part of the DDT is done by the system rather than the human driver. For active safety features, commonly referred to as "Advanced Driver Assistance Systems (ADAS)" (not to be confused with the Level 1 automation nomenclature of "Driver Assistance"), the human driver is still completing the entire DDT but the system is monitoring the driving and alerting the driver or intervening, depending on the feature. Active safety features are also identifiable by the intermittent nature of the alerts or intervention. Some common active safety features include:

- Forward collision warning (FCW)
- Lane-keeping assist (LKA)
- Blind-spot warning
- Automatic emergency braking (AEB)
- Lane-departure warning (LDW)

Some important driving automation system-related definitions from SAE J3016 (Society of Automotive Engineers, 2018) that will be used throughout this book include:

Driving Automation System: The hardware and software that are collectively capable of performing part or all of the DDT on a sustained basis; this term is used generically to describe any system capable of Level 1-5 driving automation.

Automated Driving System (ADS): The hardware and software that are collectively capable of performing the entire DDT on a sustained basis, regardless of whether it is limited to a specific operational design domain (ODD); this term is used specifically to describe a Level 3, 4, or 5 driving automation system.

Dynamic Driving Task (DDT): All of the real-time operational and tactical functions required to operate a vehicle in on-road traffic, excluding the strategic functions such as trip scheduling and selection of destinations and waypoints. The longitudinal and lateral control as well as the object and event detection and response (OEDR; see below) are included.

Operational Design Domain (ODD): The specific conditions under which a given driving automation system or feature thereof is designed to function, including, but not limited to, driving modes.

Object and Event Detection and Response (OEDR): The subtasks of the DDT that include monitoring and driving environment (detecting, recognizing, and classifying objects and events and preparing to respond as needed) and executing an appropriate response to such objects and events (i.e., as needed to complete the DDT and/or DDT fallback).

DDT Fallback: The response by the user or by an ADS to perform the DDT or achieve a minimal risk condition.

Minimal Risk Condition (MRC): The condition to which a user or an ADS may bring a vehicle after performing the DDT fallback in order to reduce the risk of a crash when a given trip cannot or should not be completed.

Scope of the Book

As discussed earlier, there are three main systems in a CAV: Perception System, Path Planning System, and Actuation System. This book will focus on the first two systems; the third is primarily a controls problem but will be briefly discussed in the context of the actuator hardware and in path planning; more information on actuation can be found in Gillespie (2021) and Kiencke and Nielsen (2005). Chapter 2 describes the various methods CAVs use to determine the location and understand the environment. Chapter 3 explores Connectivity and its impacts. Chapter 4 characterizes the various sensors used in CAVs to gather information about the environment as well as the hardware implemented for control actuation. Chapter 5 describes the computer vision techniques that are used by CAVs to process the camera-based information that is gathered. Chapter 6 outlines the sensor fusion process that takes the disparate information from the various sensors and processes the information to make sense of the environment and other entities within it. Chapter 7 covers the path planning methods for determining what trajectory the CAV will take and executing the desired trajectory. Chapter 8 describes the verification and validation (V&V) methods and processes for ensuring that a CAV is safe for deployment on public roads. Finally, Chapter 9 closes the book with discussion on what some of the next steps might be in the CAV industry, what obstacles still loom over it, and the near-, medium-, and long-term industry outlooks. Cybersecurity is not included in this introduction to CAVs; information on this topic can be found in D'Anna (2018) and Ghosh (2016).

In order to connect the chapters and provide a concrete example of a CAV "in action," Chapters 2 through 8 will open with the following driving scenario, as depicted in Figure 1.29: The CAV (in red) will be making a left turn at a four-way-stop intersection, with one car (in green) approaching the intersection from the opposite direction to the CAV, one car (in blue) stopped to the CAV's right, and a pedestrian crossing the street across the intended CAV path as it makes its left turn. A description of how the subject of the chapter pertains to this driving scenario will then provide a contextual introduction to the reader.

FIGURE 1.29 Driving scenario example.

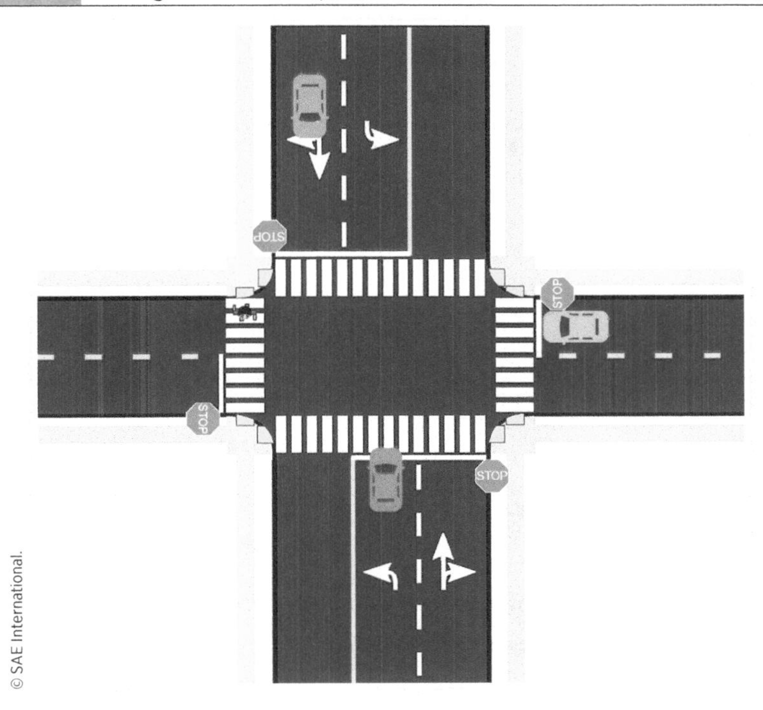

References

Baxter, J., Merced Cirino, D., Costinett, D., Tolbert, L., & Ozpineci, B. (2018). Review of Electrical Architectures and Power Requirements for Automated Vehicles. *2018 IEEE Transportation Electrification Conference and Expo (ITEC)*. Long Beach, CA. doi:10.1109/ITEC.2018.8449961

Danley's Garage World. (n.d.). Average Garage Size: What's Right for You? Retrieved from https://www.danleysgarageworld.com/average-garage-size/

D'Anna, G. (2018). *Cybersecurity for Commercial Vehicles*. SAE International, Warrendale, PA.

Ghosh, S. (2016). *Automotive Cybersecurity: From Perceived Threat to Stark Reality*. SAE International and ABOUT Publishing Group, Warrendale, PA.

Gillespie, T. (2021). *Fundamentals of Vehicle Dynamics*, 2nd Revised Edition. SAE International, Warrendale, PA.

INRIX. (2020). Global Traffic Scorecard. INRIX Research.

Joint Agency Coordination Centre. (2015). Australian Government—Department of Infrastructure, Transport, Regional Development and Communications. Retrieved from https://www.infrastructure.gov.au/aviation/joint-agency-coordination-centre/

Kiencke, U., & Nielsen, L. (2005). *Automotive Control Systems: For Engine, Driveline, and Vehicle*, 2nd Edition. Springer-Verlag, New York.

Local Motors. (n.d.). Meet Olli. Retrieved from localmotors.com: https://localmotors.com/meet-olli/

Massachusetts Institute of Technology. (2020, September 22). Moral Machine. Retrieved from https://www.moralmachine.net/

Mateja, J. (1995). Oldsmobile's $1,995 Talking Map. *Chicago Tribune*.

National Highway Traffic Safety Administration. (2018). Critical Reasons for Crashes Investigated in the National Motor Vehicle Crash Causation Survey.

Old Urbanist. (2011, December 12. We Are the 25: Looking at Street Area Percentages and Surface Parking. Retrieved from https://oldurbanist.blogspot.com/2011/12/we-are-25-looking-at-street-area.html

Puentes, R. (2017, September 18). How Commuting Is Changing. Retrieved from U.S. News & World Report: https://www.usnews.com/opinion/economic-intelligence/articles/2017-09-18/what-new-census-data-reveal-about-american-commuting-patterns

Ramachandran, S., & Stimming, U. (2015). Well-to-Wheel Analysis of Low-Carbon Alternatives for Road Traffic. *Energy & Environmental Science*, 8, 3313-3324.

Society of Automotive Engineers. (2018). J3016: Taxonomy and Definitions for Terms Related to Driving Automation Systems for On-Road Motor Vehicles.

Stephens, T., Gonder, J., Chen, Y., Lin, Z., Liu, C., & Gohlke, D. (2016). Estimated Bounds and Important Factors for Fuel Use and Consumer Costs of Connected and Automated Vehicles. U.S. Department of Energy.

U.S. Bureau of Labor Statistics. (2020). Usual Weekly Earnings of Wage and Salary Workers—Second Quarter 2020.

U.S. Bureau of Labor Statistics. (n.d.). Heavy and Tractor-trailer Truck Drivers. Retrieved from https://www.bls.gov/ooh/transportation-and-material-moving/heavy-and-tractor-trailer-truck-drivers.htm

U.S. Census Bureau. (2017a, December 7). Average One-Way Commuting Time by Metropolitan Areas. Retrieved from https://www.census.gov/library/visualizations/interactive/travel-time.html

U.S. Census Bureau. (2017b). Characteristics of New Housing.

U.S. Census Bureau. (2019, June 06). Number of Truckers at All-Time High. Retrieved from https://www.census.gov/library/stories/2019/06/america-keeps-on-trucking.html

United Nations—Department of Economic and Social Affairs. (2018, May 16). UN.org. Retrieved from https://www.un.org/development/desa/en/news/population/2018-revision-of-world-urbanization-prospects.html

Vanderbilt, T. (2012, February 6). Autonomous Cars through the Ages. *Wired Magazine.*

Weber, M. (2014, May 8). Where To? A History of Autonomous Vehicles. Retrieved from computerhistory.org: https://computerhistory.org/blog/where-to-a-history-of-autonomous-vehicles/

VOL
MAP
RADIO
MEDIA
NAV
FRONT
REAR
PASSENGER AIR BAG
A/C
CLIMATE
MODE
DUAL
CleanAir
TEMP
AUTO
TEMP

2

Localization

The first step in the perception sub-system, i.e., for the CAV to begin to understand its surroundings, is to determine its pose (also known as its 3D orientation), 3D position, speed, and acceleration and how these parameters relate to a known map. This process is known as localization. This chapter will describe the need, challenges, technologies, and techniques for this all-important perception sub-system. The characteristics and construction of the high-definition (HD) maps that are used as part of the localization process will also be discussed.

Localization can be contextualized within the driving scenario example shown in Figure 2.1. The CAV's localization algorithm will have a map of the environment that is constructed either a priori or in real time by the CAV itself (see section "Mapping"). The world coordinate system origin (X, Y, Z) of the intersection map is shown at the center of the intersection, on the surface of the road. Note that the choice of origin is arbitrary and left to the localization algorithm developer to decide. The CAV also has a local coordinate system (X′, Y′, Z′) that can be described relative to the world coordinate system. The CAV localization algorithm, using sensor data (i.e., from cameras or an inertial measurement unit [IMU]; see section "Sensing"), determines the CAV pose with respect to the local origin (note that the choice of local coordinate system alignment with the world coordinate system means that the pose will be the same with respect to both systems). Since the CAV has stopped straight in its lane, its yaw component (rotation about the Z′ axis) will be 0°; if the road is flat, then the CAV will have neither roll component (rotation about the Y′ axis) nor pitch component (rotation about the X′ axis). The localization algorithm also uses sensor data (i.e., from a GPS unit; see section "Sensing") to determine the CAV position (i.e., its local origin) with respect to the world origin. The Y and Y′ axes align, so the distance along the X axis is zero. If the road is flat (or the vehicle and origin point are at the same elevation), the relative distance from Z′ to Z is also zero. The distance from the local origin to the world origin is then along the Y and Y′ axes (i.e., a positive, nonzero value). The localization also uses sensor data (i.e., from the wheel speed sensors; see section "Sensing") to determine the CAV speed and acceleration. In this case, the CAV speed and acceleration are zero, but as the CAV completes the left turn, there will be speed and acceleration components in both the Y and X directions (again, assuming the road is flat) in addition to pitch, roll, and yaw due to the steer input and movement of the suspension. The localization determination is subsequently used in development of the world model of the CAV (see section "Sensor Fusion Definition and CAV Data Sources"). The localization input to the world model includes:

- The map of the CAV environment
- The pose, position, and motion of the CAV within the map of its environment

FIGURE 2.1 Driving scenario example.

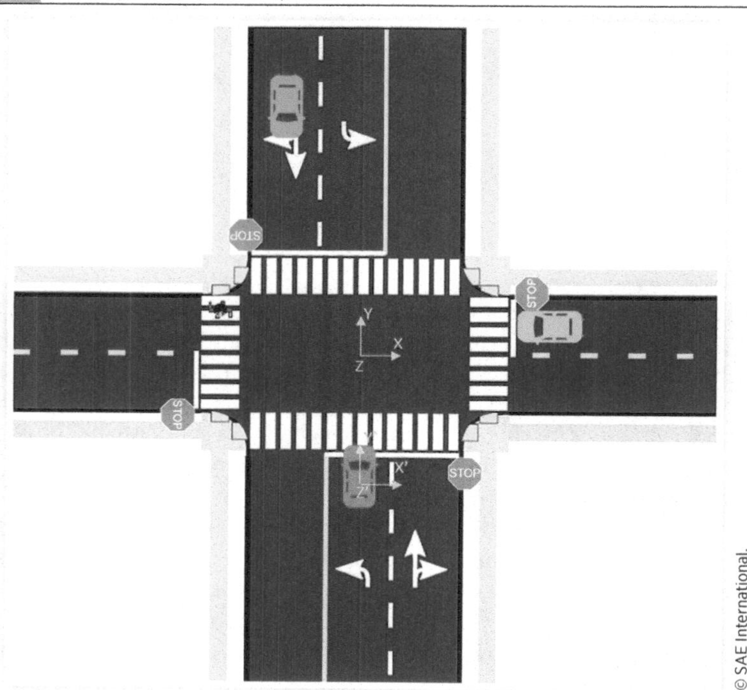

Localization Need

In order to fully perceive one's surroundings, an understanding of the current position is crucial. While a human driver may not know the GPS coordinates of his or her own vehicle precisely, a general understanding of one's location is necessary for safe navigation. ADS perception sub-systems do not faithfully recreate the perception process of humans, but instead have a perception process that is more simplistic, making localization even more important. The pose of the CAV is also crucial information to determine so that the path planning sub-system can function properly. Localization is considered a part of the perception system, as shown in Figure 2.2, and is an input to the path planning system.

FIGURE 2.2 ADS sub-systems.

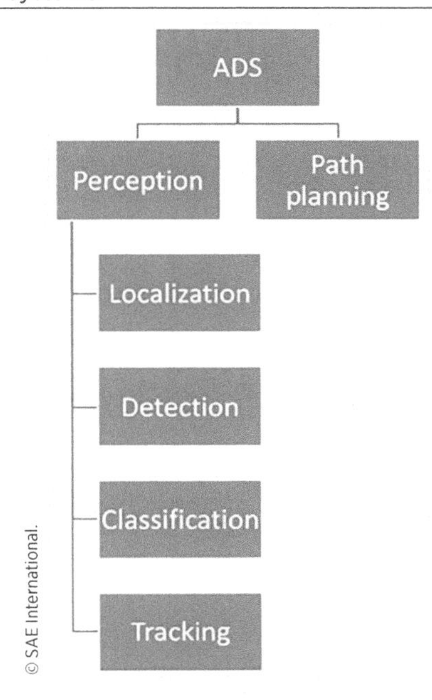

Localization can also serve as a method of perception redundancy; the localization with respect to various static entities in the map can be compared to the determinations of detected entities and the distances from these entities. Localization consists of two primary steps that will be discussed in the following sections: "Mapping" and "Sensing."

Mapping

Mapping ranges from two-dimensional (2D), conventional maps to 3D HD maps. Maps used in localization can be obtained in one of three ways:

1. Via connectivity to an outside source
2. Stored in an on-board computer
3. Created in real time

Mapping is a crucial component of localization, and many ADS developers allow their CAVs to operate with the ADS completing the DDT only in a geo-fenced area where a priori mapping has been done. In this case, the CAV has access to the map via one of the first two methods above. Real-time mapping is not generally accepted as sufficient by ADS developers, with Tesla being a major exception that will be discussed later in the chapter. The requirement for a priori mapping also means up-to-date mapping, with ADS developers updating a map if changes occur due to, for example, construction.

Many conventional maps (also known as standard definition [SD] maps) are publicly available, from such services as Google Maps, Apple Maps, Waze, etc. These maps, while useful to human drivers for navigation, are insufficiently detailed for CAVs. CAVs require details such as accurate widths of traffic lanes, heights of curbs, and clear information on traffic signs and lights. The accuracy of conventional maps is also insufficient, often displaying several meters of inaccuracy. For example, one study found that Google Earth®, the internet service that provides free access to satellite imagery, has an accuracy range of 0.4 m to 171.6 m (Potere, 2008), showing an accuracy inconsistency that is unacceptable for CAVs.

The usage of 3D HD maps is thus essential for CAVs. These 3D HD maps exhibit a high accuracy in locating objects, with a resolution at the centimeter level, and establish what is known as the "ground truth" for the ADS. If the accuracy is sufficiently high, the location of static objects can be obtained from the map rather than the sensors of the perception system in cases of sensor failure or blind spots.

With these highly detailed maps, less real-time perception and classification is required, and the ADS can focus processing power on change detection from the ground truth—identifying elements such as cars, pedestrians, etc. that would not be a component of the static map. The better the 3D HD map the less the CAV has to rely on real-time sensor data capture and processing.

The 3D HD map is constructed from a variety of sources, such as on-board vehicle sensors, satellite imagery, and a base, SD map. Vardhan (2017) proposed the following taxonomy (in "layers") for 3D HD maps used by CAVs:

- *Base map layer*—This is the SD map that includes road curvatures, elevation, and GPS coordinates.
- *Geometric map layer*—The 3D point cloud created from the mapping sensors (LIDAR, camera, GPS, etc.) that includes the stationary objects in the surroundings.
- *Semantic map layer*—2D or 3D semantic objects (such as lane markings, street signs, and other details).
- *Map priors layer*—A priori information and entity behavioral data, such as average cycle/pedestrian speeds, and SPaT information.
- *Real-time layer*—This is the real-time traffic information that a CAV can receive to assist with trip planning and navigation.

A decision must be made where to store the 3D HD map data: on board the vehicle, in the cloud, at the Network Operations Center (NOC), or at the "edge." The latency requirements of the perception step require on-board data fusion, but it is conceivable that a CAV will only store local data and will obtain the 3D HD map from off-board sources when entering a given geo-fenced zone.

When creating the maps, data can be transferred by downloading directly from the data acquisition equipment (more common in captive fleets) or from vehicle to cloud and cloud to central database (more common in crowdsourcing).

Sensing

In the context of localization, sensing refers to the use of sensors to obtain information that allows for the position and heading of the subject CAV to be determined relative to the 3D HD map. Further details on sensing in the context of CAV sensors that are used to perceive other road users, traffic lanes, static objects, etc. in the CAV environment are included in Chapter 4. The CAV's sensing must be functional and accurate in a variety of conditions, such as low lighting, in tunnels, among tall buildings and trees, and in varying weather conditions. The data from the various sensors will all be in different formats and must be "fused" together through sensor fusion methods which will be discussed in Chapter 6.

The sensor technologies used in localization can be either "exteroceptive" or "proprioceptive". Exteroceptive sensors are used in both mapping and sensing aspects of localization, while proprioceptive sensors are used exclusively in sensing:

- *Exteroceptive*: External to the vehicle, i.e., from the vehicle surroundings
- *Proprioceptive*: Internal to the vehicle, i.e., from the vehicle itself

The most popular exteroceptive sensors currently in use are described below, although Chapter 4 provides a more in-depth exploration into each of these sensor types:

- Global Navigation Satellite Systems (GNSS): GNSS (in the United States, the most common GNSS is the GPS that uses the NAVSTAR satellite network developed by the U.S. Department of Defense) sensors use satellite signals to obtain subject CAV coordinates on the ground. These coordinates can then be compared to the coordinates on the 3D HD map. The nominal accuracy is 1-10 m, which is insufficient for CAV localization. The lack of accuracy of GNSS is primarily due to imprecision in satellite orbit, satellite clock errors, and atmospheric disturbances. The GNSS accuracy can be improved by using a "correction service" in which reference stations (sometimes referred to as "base stations" for differential GPS units) with accurately known locations receive the GNSS coordinates and calculate the required correction.
- LIght Detection And Ranging (LIDAR): LIDAR sensors are used in localization for 3D HD map creation, specifically for measuring the distances between entities in the surroundings to the subject CAV, and also during the sensing phase for the same distance calculations that can be compared against the 3D HD map.
- RAdio Detection And Ranging (RADAR): RADAR sensors are less commonly used in localization, but when they are used, it is during the sensing phase, specifically for measuring the speeds of entities in the surroundings to the subject CAV to ensure that identified stationary objects are indeed stationary.
- Camera: Camera sensors are used in localization for 3D HD map creation, specifically for semantic identification of the entities in the surroundings to the subject CAV, and also during the sensing phase for the same semantic identifications that can be compared against the 3D HD map.
- Ultrasonic Sound Navigation and Ranging (SONAR): Ultrasonic SONAR sensors are used for detection and ranging at small distances (within 5 m) near the subject CAV. Ultrasonic SONAR units are cheaper and require less processing than RADAR and LIDAR signals, making the former an attractive modality in a CAV sensor array. Ultrasonic SONAR also has high directivity, which is useful in situations that require precision, such as parking. Parking is the main application for ultrasonic SONAR in CAVs, with ultrasonic SONAR sensors installed with overlapping fields of view (FOVs) in the front and rear bumpers.

Proprioceptive sensors include:

- Inertial Measurement Unit (IMU): IMU sensors use accelerometers to measure acceleration in the X-Y-Z directions, but also gyroscopes to measure rotation about the roll, pitch, and yaw principal axes. To measure movement from a specific location, IMU sensors are mounted in a fixed position in the subject vehicle. IMUs can be thought of as analogous to the inner ear in a human that helps with balance and perception of motion.

- Wheel Speed Sensor: Wheel speed sensors are used in odometry to measure the change in position of the subject vehicle over a set period of time, i.e., the vehicle speed. The sensors are usually an optical encoder that uses a light source and optical detector; the number of times that the light source is detected corresponds to the number of wheel revolutions and speed of revolution. Having encoders on all four wheels for speed differences of the wheels, for example, allows for subject vehicle turns to be captured although this can also lead to inconsistencies in vehicle speed due to wheel slippage and events causing wheels to leave the ground plane.
- Steering Angle and Torque Sensors: Steering angle and torque sensors can also be used in odometry, incorporating turns in the distance and velocity determinations. Steering angle sensors can be analog or digital and can be connected to the electronic stability control (ESC) (ESC was required in all new vehicles in the United States starting with the 2012 model year) (Delphi Technologies, n.d.).

As in all data collection, calibration of sensors is critical to ensure accuracy. This is especially true in localization, where centimeter resolution is required. An example of a required calibration would be the case of multiple LIDAR units being employed, where the sensors' responses (i.e., reflected laser beams and timing) to the same object must be the same, as shown in Figure 2.3. The bottom figure shows that uniform reflectivity from the same object makes for a clearer view of the object.

FIGURE 2.3 Intensity calibration of a LIDAR sensor.

(a) Uncalibrated sensor

(b) After intensity calibration

Localization Challenges

There are myriad challenges associated with CAV localization, ranging from logistical to technical. For the mapping portion of localization, the main challenge is how to create a 3D HD map that is sufficiently detailed over a wide area in order to be useful. The map must exhibit a high degree of accuracy such that the CAV can treat the map as ground truth and conduct the localization with respect to the ground truth. The area of coverage must be widespread or the ODD of the CAV will be limited. Although road infrastructure is not in constant flux, changes are made periodically, such as during construction. Other examples include lane line fading and sign deterioration/damage. These changes must all be captured; therefore, the 3D HD map must be maintained so that it is always up to date.

The question of how the 3D HD map is created is another challenge. There are two main strategies currently being used:

1. Mapping services organization
2. Crowdsourcing organization

A mapping services organization has its own captive vehicle fleet deployed with map data collection hardware. The fleet vehicles are driven through a desired area with the express purpose of collecting data. The main advantage enjoyed by such a strategy is that this is an entity with dedicated resources to accurate data collection and 3D HD map generation. The main disadvantage is that the data collection is expensive, and the fleet must be large to cover an extensive area. To provide context on the cost, in 2018 a mapping services company DeepMap, a fleet vehicle of which is shown in Figure 2.4, was reportedly charging $5,000 USD per kilometer for its mapping service [techcrunch.com, DeepMap]. As mentioned above, changes to the road infrastructure must be captured; thus, the mapping services organization must return to locations where change has occurred. If information on where changes have occurred is available, the return mapping trips can be targeted. However, it is unlikely that every change will be known such that targeted mapping can always occur; more likely is that some portion of a mapped area must have returned mapping periodically to account for unknown changes. Balancing the need for an up-to-date 3D HD map and the cost of creating the map will always be a major consideration for the mapping services organization.

FIGURE 2.4 DeepMap data collection fleet vehicle.

Using a crowdsourcing strategy for 3D HD map creation involves using non-fleet vehicles for the data capture. In this case, a crowdsourcing map organization provides data collection technology to vehicles owned by others (e.g., members of the general public who agree to participate, perhaps for some form of compensation) and then uses these data in the map generation. There are several variants to the crowdsourcing model. A company called Lv15, for example, pays drivers to use their own iPhones mounted as dashcams in their vehicles along with Lv15's phone application Payver to accumulate data with a compensation of 0.05 USD per mile (Kouri, 2018). Lv15 is working with Uber and Lyft to scale their data collection. Conversely, Mapbox is working with active safety systems and ADS developers to collect data from deployed vehicles and prototype testing, respectively. The sensors allow for better data collection than using a smartphone camera, but the scaling of the enterprise is more difficult to achieve.

The main advantages of using crowdsourcing include low cost and fast data accumulation. Crowdsourcing also allows for updates of changes in maps to happen without requiring knowledge of the changes. However, the main disadvantages include generally lower-quality data when limited sensors are used, so the data need to be filtered and post-processed. Furthermore, since the vehicle fleets are not captive, the area coverage cannot be guaranteed, so there may be gaps. Further, outsourcing the data collection means that contracts with entities with varying interests must be signed, and the mapping company does not control the data collection even though it must, in some cases, provide the hardware.

The common theme of both strategies is that cost of 3D HD map development is a large—perhaps the largest—consideration. Which strategy proves more feasible and how the required 3D HD maps are developed remain open questions. Another unknown is whether governments and jurisdictional authorities will develop (and maintain) these maps because they are deemed a public good and provide them to the CAV community.

One paradigm that can mitigate the need for a CAV to have access to a single, comprehensive 3D HD map is to employ a vehicular ad hoc network (VANET). In this paradigm, a CAV will be connected with other CAVs and/or roadside units (RSUs), each with some or all of the local 3D HD map that can be shared with the other connections. Figure 2.5 depicts how the VANET and the 3D HD maps can be combined with real-time traffic data for better localization.

FIGURE 2.5 VANET for 3D HD map sharing.

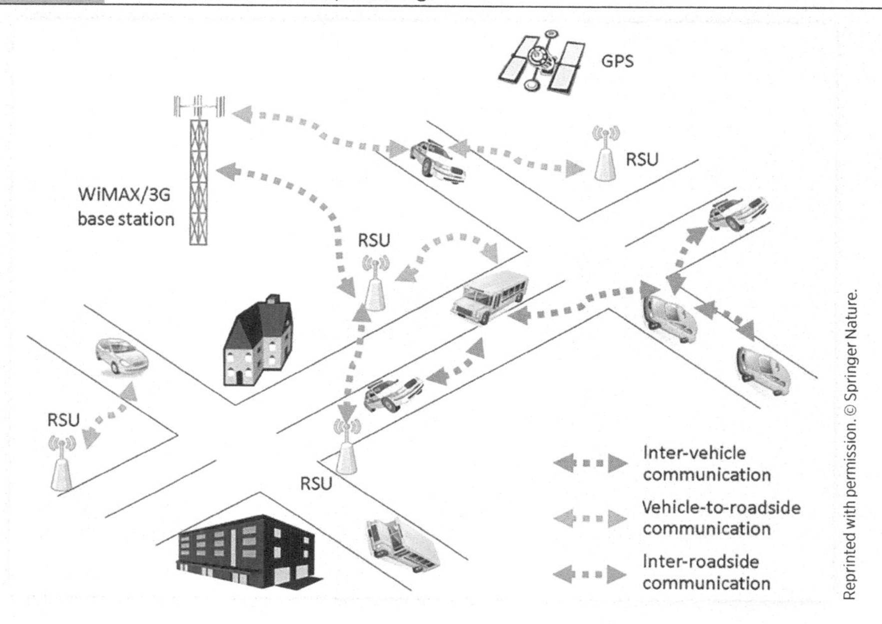

Other localization challenges include map data challenges. For example, the data will have associated levels of uncertainty, just as any data do. The usefulness and ability of a CAV to rely on the map data is highly dependent upon an understanding of the level of uncertainty and a minimization of the uncertainty wherever possible. The main areas of uncertainty are:

1. *Existence:* Does an object in the mapping data actually exist?
2. *Accuracy:* How accurate is the location of the object in the mapping data?
3. *Classification:* Is the classification of an object in the mapping data accurate?

It is helpful to have knowledge of the sensors used in the 3D HD map creation and their limitations in order to understand these uncertainties.

Finally, a major localization challenge is the trade-off between real-time requirements and computational complexity. For localization to be useful, it must be done in real time, and while a more complex 3D HD map may contain useful information, it makes the real-time usage more difficult. Techniques, such as those described in the following section, exist to mitigate this trade-off while maintaining the needed accuracy of localization.

Localization Techniques

The 3D HD map is normally comprised of a fixed set of elements and representations within an environment. In this case, the binary classification that results from interpretation of sensor data is as either a true object or spurious data. However, the map can also be viewed as a series of probability distributions as shown in Figure 2.6. Thus a probability can be assigned to detected objects by calculating an average and variance of the measurements. This technique can make the localization more robust to environment changes, more accurate, and help avoid localization failures such as false positives and false negatives.

FIGURE 2.6 Probability distribution technique for 3D HD map development.

(a) Average infrared reflectivity

(b) Standard deviation of infrared reflectivity values

Other localization techniques include ones that allow for localization to be performed even in the absence of a pre-built 3D HD map, or to be accurate when elements of the existing, pre-built 3D HD map have changed, and are known as Simultaneous Localization and Mapping (SLAM) algorithms. SLAM algorithms were first developed in the 1980s and are a core part of robotics. SLAM algorithms require more processing power and can be less accurate but can perform the localization in real time. SLAM algorithms result in incremental map construction as the subject vehicle moves through the surroundings and simultaneously performs localization. In terms of automotive manufacturers developing CAVs, Tesla is the most prominent proponent of the use of SLAM. At the time of writing, Tesla does not pre-build 3D HD maps, and its vehicles using the Autopilot driving automation system are employing SLAM algorithms.

One possible method of SLAM is via implementation of a Bayes filter, in particular the particle filter technique, where particles are created throughout the location space and weighted by an amount that indicates the discrete probability that the vehicle is at that location. The particle filter algorithm has four stages, as shown in Figure 2.7.

FIGURE 2.7 Particle filter flowchart.

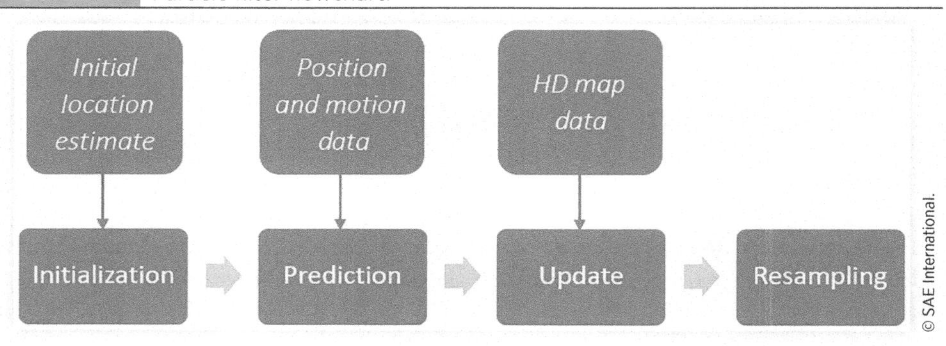

1. Initialization—Take an initial estimate from GPS to obtain a distribution of possible locations (i.e., particles) with equal weights within the GPS space where the vehicle could be located. Each particle has a position (x, y) and orientation θ.
2. Prediction—Taking into account the data from the subject CAV sensors, e.g., the IMU(s) (such as longitudinal/lateral velocity and acceleration), a prediction is made on the position or motion of the vehicle and weights of the distribution. If IMU sensors are used and motion data are used, the equations of motion that determine the final position (x_f, y_f) and orientation θ_f based on the initial position (x_i, y_i) and orientation θ_i are (Cohen, 2018):

$$x_f = x_0 + \frac{v}{\dot{\theta}}\left[\sin\left(\theta_0 + \dot{\theta}(dt)\right) - \sin\left(\theta_0\right)\right] \quad (1)$$

$$y_f = y_0 + \frac{v}{\dot{\theta}}\left[\cos\left(\theta_0\right) - \cos\left(\theta_0 + \dot{\theta}(dt)\right)\right] \quad (2)$$

$$\theta_f = \theta_0 + \dot{\theta}(dt) \quad (3)$$

where v is the velocity measured by the IMU.

3. Update—The match between measurements of location and the 3D HD map is analyzed, and the weights assigned to the possible locations are modified according to the equation (Cohen, 2018):

$$P(x,y) = \frac{1}{2\pi\sigma_x\sigma_y} e^{-\left(\frac{(x-\mu_x)^2}{2\sigma_x^2} + \frac{(y-\mu_y)^2}{2\sigma_y^2}\right)} \quad (4)$$

Here σ_x and σ_y are the uncertainties of the measurements, x and y are the observations of the landmarks, and μ_x and μ_y are the ground truth coordinates of the landmarks derived from the 3D HD map. When the error of the predicted location is large, the exponent in the equation is 0 and the probability (weight) is 0. When the error is small, the weight is 1 (standardized by the denominator).

4. Resampling—Only the locations with the highest weights are kept; the locations with the lowest weights are discarded.

The cycle is repeated with the most likely locations until a final location is determined. Note that the sensors associated with the data inputs to the Initialization, Prediction, and Update steps will all have different types of errors and noise for which the algorithm needs to account. A depiction of the particle filter technique using LIDAR measurements is shown in Figure 2.8. Here the LIDAR measurements (in red) are compared against the map landmarks (in blue) as described in Step 3 above.

FIGURE 2.8 Particle filter technique depicted with LIDAR sensor measurements.

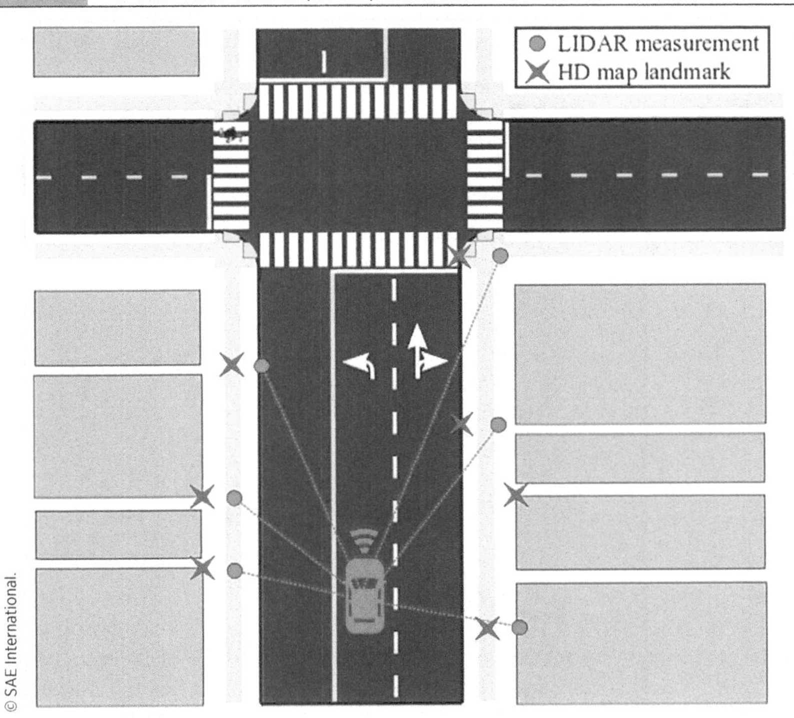

Other techniques for SLAM exist, and another common type of filter is the Kalman (Extended Kalman Filter or Unscented Kalman Filter) filter. SLAM can also be accomplished through optimization (i.e., Graph SLAM), where the vehicle's pose (derived from sensor measurements) imposes constraints on the vehicle motion with respect to the map and landmark information. The optimal cost function of the poses is a nonlinear optimization problem to be solved. Deep learning-based methods such as Convolutional Neural Network-SLAM are also being developed.

The common problems with SLAM algorithm results, illustrated in Figure 2.9, include:

- Accumulation of errors, in which the localization worsens as the CAV proceeds
- Robustness of the algorithms to new types of environments and faults
- Computational cost meaning that CAVs will expend a significant proportion of the computing power on localization, which reduces the computational power left over for perception, path planning, and actuation

FIGURE 2.9 SLAM results illustration.

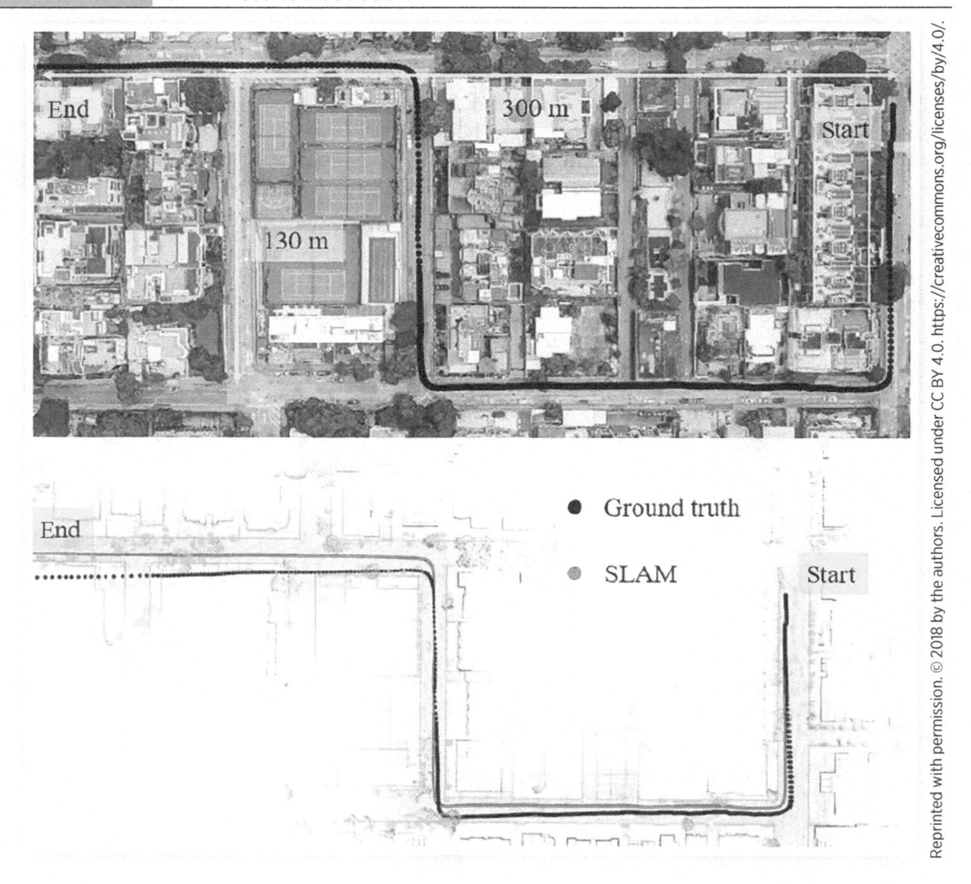

This chapter has outlined the localization step in the CAV's completion of the DDT. Localization is a key component for the CAV's understanding of its environment, and deviation from ground truth can lead to errors and potential unsafe driving behavior. As such, localization is critical to proper CAV functionality. Additionally, accurate localization can be shared with other traffic participants in the environment for overall improvement in network safety, as will be discussed in the next chapter.

References

Cohen, J. (2018, June 26). Self-Driving Cars & Localization. Retrieved from towardsdatascience.com: towardsdatascience.com/self-driving-car-localization-f800d4d8da49

Delphi Technologies. (n.d.). Making Sense of Sensors: Steering Angle Sensor. Retrieved from https://www.delphiautoparts.com/usa/en-US/resource-center/making-sense-sensors-steering-angle-sensor

Kouri, A. (2018, October 30). Introducing Crowdsourced Pavement Quality Maps. Retrieved from medium.com: https://medium.com/lvl5/introducing-crowdsourced-pavement-quality-maps-8ddafd15a903

Levinson, J., & Thrun, S. (2010). Robust Vehicle Localization in Urban Environments Using Probabilistic Maps. *IEEE International Conference on Robotics and Automation*. Anchorage, AK.

Lunden, I. (2018, November 1). DeepMap, a Maker of HD Maps for Self-Driving, Raised at least $60M at a $450M Valuation. Retrieved from techcrunch.com: techcrunch.com/2018/11/01/deepmap-a-maker-of-hd-maps-for-self-driving-raised-at-least-60m-at-a-450m-valuation/

Middlesex University London. (n.d.). Intelligent Transport Systems (ITS). Retrieved from Vehicular Ad Hoc Network (VANET): vanet.mac.uk

Potere, D. (2008). Horizontal Positional Accuracy of Google Earth's High-Resolution Imagery Archive. *Sensors (Basel)*, 8(12), 7973-7981.

Vardhan, H. (2017, September 22). HD Maps: New Age Maps Powering Autonomous Vehicles. Retrieved from Geospatial World: https://www.geospatialworld.net/article/hd-maps-autonomous-vehicles/

Wen, W., Hsu, L.-T., & Zhang, G. (2018). Performance Analysis of NDT-Based Graph SLAM for Autonomous Vehicle in Diverse Typical Driving Scenarios in Hong Kong. *Sensors*, 18(3928), 3928. doi:10.3390/s18113928

VOL
MAP
RADIO
MEDIA
NAV
FRONT
REAR
PASSENGER
AIR BAG
A/C
CLIMATE
MODE
DUAL
CleanAir
TEMP
AUTO
TEMP

3

Connectivity

While connectivity may not receive as much attention as automation by either the media or the public, this technology could be a key enabler to CAVs becoming widespread on public roads. Connectivity is a complement to both the localization and perception sub-systems. Information can be obtained from off-board sources that can enhance, provide redundancy for, or even correct the values from these other subsystems so that the CAV has a more accurate understanding of its surroundings.

The importance of connectivity to the safe introduction of automated vehicles can be demonstrated using the driving scenario example shown in Figure 3.1. In the scenario, the three CAVs, as well as the pedestrian, are broadcasting data about their position and movement that can be received by other road users in the environment. This allows the three vehicles to know that a pedestrian is crossing the street from north to south, which increases the awareness of their location and motion in case they are not visible due to inclement weather, lighting conditions, or even a sensor failure on a CAV. The increased awareness of the presence of the pedestrian allows the red vehicle making the left turn to delay entering the intersection until the pedestrian has completed the crossing. This delay reduces the likelihood of an accident, and thus connectivity has improved the traffic safety of this scenario. In the case of a sensor failure on a CAV, the connectivity has provided safety redundancy for the automation. The CAVs could also be cooperating, and in addition to sharing data on position, velocity, acceleration, and even intent, the ADS could be agreeing to particular trajectories dictated by another vehicle or even an intersection manager.

FIGURE 3.1 Driving scenario example.

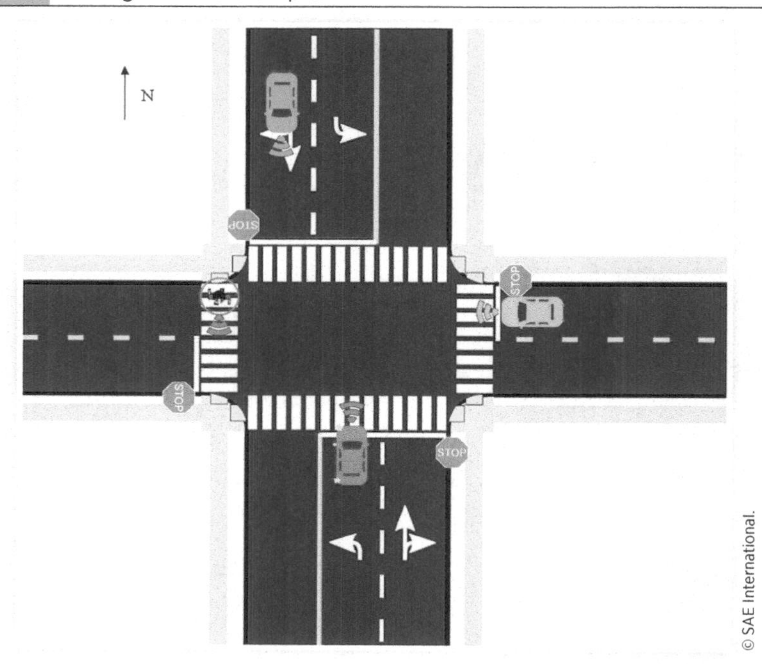

This chapter will cover the definition of connectivity as used in this book. The history of connectivity in the automotive sector will then be summarized. The motivations behind why connectivity is seen as such an important part of the future of the automotive sector will then be discussed. The technologies behind connectivity will be covered, including the competing Dedicated Short-Range Communication (DSRC) and fifth-generation (5G) technologies. Finally, the deployment challenges and potential impacts will be examined.

Connectivity Defined

Connectivity, in the context used in this book, is technology that enables information to be transferred to and from the CAV to off-board sources such as other vehicles, bicycles, pedestrians, and the infrastructure, as depicted in Figure 3.2. Infrastructure connectivity could mean information sharing with the cloud, an NOC, or a GPS ground node. In this context, any information that would not otherwise be available but that is transferred to or from the vehicle constitutes connectivity, for example, navigation maps that are not stored on board the vehicle (even if the connection is enabled by a smartphone in the vehicle). Connectivity allows for some decisions to be made by the CAV collaboratively rather than exclusively individually. It also allows for better decisions to be made due to additional information. This information permits vehicles to have significantly expanded digital "line-of-sight" capabilities from on-board sensors that will be discussed in Chapter 4, allowing for "seeing" around corners and through other vehicles.

FIGURE 3.2 Depiction of connectivity.

Reprinted from Public Domain.

The three main types of CAV connectivity are:

- Vehicle-to-Vehicle (V2V)
- Vehicle-to-Infrastructure (V2I)
- Vehicle-to-Pedestrian (V2P)

Along with Vehicle-to-Grid (V2G) and Vehicle-to-Device (V2D), collectively, these types of connectivity are known as Vehicle-to-Everything (V2X). In V2X connectivity, information is shared among entities within a given defined area, including:

- Speed
- Location
- Heading
- Brake status
- Traffic SPaT
- Traffic status
- Traffic hazards, including construction
- Lane locations and rules (e.g., travel direction)

The speed, location, and heading information are of particular note as these parameters are part of a list of so-called Basic Safety Messages (BSMs) that were designated in the SAE J2735 standard. Standardization of messages is essential so that all participating connected entities are speaking the same "language." The traffic SPaT, status, hazards, and lane locations are information parameters that are considered to be part of an analogous list for infrastructure, or Basic Infrastructure Message (BIM). However, there is not yet a standard for BIMs. The SAE J2735 document (first published in 2006 and updated several times, including the latest update in 2020) establishes a standard set of messages, data elements, and data frames in the general V2X case. These messages, which can be broadcast and either acknowledged or not by the receiver, include information relating to the surrounding environment such as dangers of nearby vehicles disregarding right-of-way laws, emergency vehicle presence, and other data regarding the driving environment. The complete list of BSMs from SAE J2735 is included in Appendix A (SAE International, 2020):

1. MSG_MessageFrame (FRAME): This message contains all the defined messages of the SAE J2735 standard.
2. MSG_BasicSafetyMessage (BSM): Safety-related information broadcast periodically by and to the surrounding traffic participants and road operators (transmission rates are not in the standard's scope, but 10 Hz is said to be typical).
3. MSG_CommonSafetyRequest (CSR): A message that one traffic participant can transmit to the others to request additional safety-related information.
4. MSG_EmergencyVehicleAlert (EVA): A warning message to traffic participants and road operators that an emergency vehicle is in the vicinity.
5. MSG_IntersectionCollisionAvoidance (ICA): A warning message to traffic participants that a potential collision is imminent due to a vehicle entering the intersection without the right of way.
6. MSG_MapData (MAP): Message containing a variety of geographic road information, such as road geometry (including curve radii) and intersection descriptions.
7. MSG_NMEAcorrections (NMEA): A message to incorporate National Marine Electronics Association (NMEA) differential corrections for GPS/GNSS signals.
8. MSG_PersonalSafetyMessage (PSM): A message to broadcast the kinematic states of vulnerable road users (VRUs).
9. MSG_ProbeDataManagement (PDM): A message for controlling the type of data transmitted between On-Board Units (OBUs) and RSU.
10. MSG_ProbeVehicleData (PVD): A message on the status of a vehicle in the traffic environment to allow for collection of vehicle behaviors along a particular road segment.
11. MSG_RoadSideAlert (RSA): A warning message sent to traffic participants in the vicinity of nearby hazards such as "slippery conditions" or "oncoming train".
12. MSG_RTCMcorrections (RTCM): A message to incorporate Radio Technical Commission for Maritime Services (RTCM) different corrections for GPS and other radio navigation signals.
13. MSG_SignalPhaseAndTiming Message (SPaT): A message to indicate the SPaT of a nearby signalized intersection.
14. MSG_SignalRequestMessage (SRM): A message sent to the RSU of a signalized intersection for either a priority signal request or preemption signal request and can include time of arrival and duration of the request.
15. MSG_SignalStatusMessage (SSM): A status message transmitted by the RSU of a signalized intersection that includes the pending and active priority and preemption requests that have been acknowledged.
16. MSG_TravelerInformationMessage (TIM): A broad message type for transmitting various advisory types of information for traffic participants and can include local place names.
17. MSG_TestMessages: Expandable message type for local road operators to provide region-specific information.

Connectivity and automation are combined in what is known as cooperative driving automation (CDA). In 2020, SAE International released J3216_202005, which outlined the various aspects of CDA systems. The classes of CDA can be mapped against the levels of automation according to SAE J3016_201806, as shown in Figure 3.3.

FIGURE 3.3 SAE J3216_202005 table mapping levels of CDA versus levels of automation.

	SAE Driving Automation Levels					
	No Automation	Driving Automation System		Automated Driving System (ADS)		
	Level 0 *No Driving Automation (human does all driving)*	Level 1 *Driver Assistance (longitudinal OR lateral vehicle motion control)*	Level 2 *Partial Driving Automation (longitudinal AND lateral vehicle motion control)*	Level 3 *Conditional Driving Automation*	Level 4 *High Driving Automation*	Level 5 *Full Driving Automation*
No Cooperative Automation	(e.g. Signage, TCD)	Relies on driver to complete the DDT and to supervise feature performance in real-time.		Relies on ADS to perform complete DDT under defined conditions (fallback condition performance varies between levels)		
CDA Cooperation Classes						
Class A: Status-sharing *Here I am and here is what I see*	(e.g., Brake Lights, Traffic Signal)	Limited cooperation: Human is driving and must supervise CDA features (and may intervene at any time), and sensing capabilities may be limited compared to C-ADS.		C-ADS has full authority to decide actions. Improved C-ADS situational awareness beyond on-board sensing capabilities and increased awareness of C-ADS state by surrounding road users and operators		
Class B: Intent-Sharing *This is what I plan to do*	(e.g., Turn Signal, Merge)	Limited cooperation (only longitudinal OR lateral intent that may be overridden by human)	Limited cooperation (both longitudinal AND lateral intent that may be overridden by human)	C-ADS has full authority to decide actions. Improved C-ADS situational awareness through increased prediction reliability, and increased awareness of C-ADS plans by surrounding road users and operators.		
Class C: Agreement-Seeking *Let's do this together*	(e.g., Hand Signals, Merge)	N/A	N/A	C-ADS has full authority to decide actions. Improved ability of C-ADS and transportation system to attain mutual goals by accepting or suggesting actions in coordination with surrounding road users and operators.		
Class D: Prescriptive *I will do as directed*	(e.g., Hand Signals, Lane Assignment by Officials)	N/A	N/A	C-ADS has full authority to decide actions except for very specific circumstances in which it is designed to accept and adhere to a prescriptive communication.		

The classes of CDA include (SAE International Surface Vehicle Information Report (J3216), 2020):

- Class A—Status-sharing: Traffic participants share their locations as well as information related to how the traffic participant has perceived the traffic environment.
- Class B—Intent-sharing: Traffic participants share their future, planned actions.
- Class C—Agreement-seeking: Traffic participants engage in a collaborative decision process for future, planned actions.
- Class D—Prescriptive: Traffic participants receive specific directions regarding future, planned actions.

It should be noted that for Class D, the directions could be originating from a CAV on the road, but also from the Infrastructure Owner-Operator in the form of an Intersection Manager (IM). IMs could conceivably allow for the removal of traffic signs and traffic lights, with CDA allowing for seamless navigation for all CAVs that greatly increase intersection throughput.

Connectivity Origins

Connectivity as a concept in the automotive industry has a long history, but without large successes in deployment beyond demonstrations. The obstacles have included the cost of the technology and the lack of a clear net benefit, whether those costs are borne by consumers of vehicles or infrastructure owners. The potential impacts on safety, mobility, and efficiency have been known for quite some time, and the concept appears to be gaining momentum.

The U.S. company GM was the first OEM to bring a connected vehicle to market with their OnStar program in 1996. OnStar was originally a partnership between GM and Motorola Automotive (the latter was bought by Continental). The original intent of OnStar was to provide emergency help in the event of an accident. The first deployment included only voice, so the driver could speak with an OnStar representative. Later, when data were added, the GPS location of the vehicle could be sent to the NOC, which helped when the driver could not verbally communicate with the automatic crash response feature.

By 2003, additional features included in OnStar are:

- Stolen Vehicle Assistance
- Turn-by-Turn Navigation
- Hands-Free Calling
- Bluetooth
- Remote Diagnostics

Also in 2003, the U.S. DOT convened the Vehicle Infrastructure Integration program with the American Association of State Highway and Transportation Officials (AASHTO) as well as OEMs to conduct research and development and deployment of connectivity technologies. Originally, V2V was seen as necessary to maximize benefits, but costs and slow fleet turnover resulted in the view that V2I could be an interim solution to achieve goals more quickly.

Subsequently, internet access was introduced by various OEMs in their vehicles. In 2004, BMW introduced built-in SIM cards in a vehicle to allow internet access. In 2005, Audi began the development of its "Audi connect" services, which were introduced in vehicles in 2009. The service brought mobile broadband internet access to vehicles. Chrysler was the first to introduce a Hotspot feature in 2009 to allow users to connect the smartphones that had been introduced at that time, most notably the Apple iPhone in 2008.

In 2011, the U.S. DOT began the Connected Vehicle Safety Pilot Program. This was a real-world implementation of connectivity technology with the public used as drivers. The program tested performance, human factors, and usability; evaluated policies and processes; and gathered empirical data. Driver clinics were held in 6 U.S. cities with 100 drivers and 3,000 vehicles equipped with connectivity technologies used to test safety applications. The results were compiled by the NHTSA to develop a communication guideline that was released in 2014.

Also in 2011, the U.S. DOT helped fund research testbeds. One such example is a collaboration between the SMARTDrive Maricopa County Department of Transportation (MCDOT) and the University of Arizona (U of A) to create the SMARTDrive Connected Vehicle testbed in Anthem, AZ. (Maricopa County Department of Transportation, 2017) Maricopa County has invested $1.1M and the federal government has invested $2M through the lifetime of the SMARTDrive project to date while U of A contributes personnel and expertise. SMARTDrive objectives include securing additional federal funding for installation of CAV-enabling technology in 3,000 Anthem residents' vehicles, plus school buses and emergency vehicles. Technology being tested includes:

- SPaT priority request to allow emergency vehicles to have green lights at intersections but also allow for greater long-haul trucking efficiency at times of very low traffic (i.e., late at night).
- V2P so that pedestrians that have disabilities or other impediments can cross the street safely.
- Red light warning to drivers to indicate that they should slow down, or failing that, SPaT control to extend the green light.

In 2014, Audi was the first OEM to offer 4G LTE hotspots in the A3 model. The hotspot could support up to 8 devices, although there were concerns among customers about data costs (Howard, 2014).

The U.S. DOT Connected Vehicle Deployment program began in September 2015 with testbeds in Wyoming, New York City, and Tampa, FL (a diverse selection of locations). Also in 2015, the U.S. DOT Smart City Challenge was initiated. This program challenged midsized cities across the United States to share ideas on how to create a transportation system that uses data, applications, and technology to move goods faster, cheaper, and more efficiently. Some 78 cities responded with ideas, including (US Department of Transportation, 2016) (quoted):

San Francisco

- GOAL: Grow the number of regional commuters that use carpooling to improve affordability, increase mobility and relieve congestion on roads and transit.
- SOLUTION:
  - Create connected regional carpool lanes and designate curb space for carpool pickup/drop off.
  - Make carpooling easy by developing a smartphone app for instant carpool matching and establish carpool pickup plazas for riders without smart phones.
  - Use connected infrastructure to monitor and optimize the performance of carpool lanes.

Denver

- GOAL: Make freight delivery more reliable and reduce air pollution, idling, and engine noise
- SOLUTION: Establish a connected freight efficiency corridor with comprehensive freight parking and traffic information systems, freight signal prioritization, designated parking and staging areas.

Pittsburgh

- GOAL: Jump-start electric conversion to reduce transportation emissions by 50% by 2030 through demonstration projects in street-lighting, electric vehicles and power generation

- SOLUTION:
  - Convert up to 40,000 street lights to LEDs to reduce energy usage
  - Establish smart street lights with sensors to monitor local air quality
  - Install electric vehicle charging stations
  - Convert the city's public fleet to electric vehicles

Kansas City

- GOAL: Advance our understanding of urban travel and quality of life to inform the transportation decisions of citizens and public officials
- SOLUTION: Make the urban core a more "Quantified Community" by collecting and analyzing data on travel flows, traffic crashes, energy usage, air pollution, residents' health and physical activity.

Also in 2015, Toyota launched a DSRC-based V2V system in Japan in which information such as vehicle speeds are shared. By March 2018, there were over 100,000 DSRC-equipped Toyota and Lexus vehicles on Japanese roads (Slovick, 2018).

In 2016, Audi launched a pilot V2I Connected Vehicle (CV) project using Fourth Generation Long Term Evolution (4G LTE) in Las Vegas, NV, in which SPaT information, including "time-to-green," is provided to some A4 and Q7 vehicles starting in model year (MY) 2017. Information is provided either on the instrument cluster or in a heads-up display (HUD, if equipped).

Since launch, Audi has expanded the project to include (Walz, 2020):

- Las Vegas, NV
- Portland, OR
- Palo Alto and Arcadia, CA
- Washington, D.C.
- Kansas City, MO
- Dallas and Houston, TX
- Phoenix, AZ
- Denver, CO
- New York, NY
- Orlando, FL

The number of intersections has grown as well, with 2,000 in New York City and 1,600 in Washington, D.C. Audi has also expanded to the German cities of Düsseldorf and Ingolstadt. The connectivity technology has also been advanced, including integration with the start-stop function, Green Light Optimized Speed Advisory (GLOSA), and optimized navigation routing.

GM launched the first V2V system in the United States in 2017, and it is now a standard feature on Cadillac CTS sedans (Cadillac Customer Experience, 2017). The feature alerts drivers to potential hazards such as hard braking, slippery conditions, and disabled vehicles.

The connected vehicle history in the United States does not become richer after the GM Cadillac CTS deployment. Toyota announced plans to equip its entire fleet with connected vehicle technology in 2018, but paused this deployment in 2019 (Shepardson, 2019). Outside of pilot projects, very few connected vehicles have been deployed commercially. It is possible that the automotive industry was not convinced that the technology was sufficiently developed or that the federal government, and specifically the U.S. Federal Communications Commission (FCC) and NHTSA agencies, was committed to supporting the current technology (as will be discussed later in this chapter) (White, 2020; Jaillet, 2013). It would be fair to say that as of this writing (mid-2021) the previous decade has been one of disappointment for the connected vehicle industry. However, there is continued and renewed interest among various stakeholders in the transportation industry, and connected vehicle technology has the potential to be a major influence and factor in the coming years.

Motivations: The Case for Connectivity

There are limits to the efficacy of using on-board vehicle sensors that allows the CAV to develop what is known as the world model (more discussion on this topic in Chapter 6—Sensor Fusion), or "understanding" of the surrounding environment. These limits include:

- The perception environment of the sensors cannot exceed the sensing range.
- The sensor systems cannot perform well in all environments: urban settings and inclement weather pose serious challenges.
- The system latency is sometimes too large in order to warn the driver (Levels 0-3) or avoid an accident (Levels 4 and 5).
- The cost of the sensor hardware and sensor fusion systems is high and will be available to non-luxury vehicles slowly.
- Storage of HD maps on board requires large data storage, which is expensive. Connectivity allows for proximal HD map download (although the bandwidth and speed of the connection must be high).
- Redundancy is increased because the world model is less reliant on any single sensor that can fail or be obscured.

The limitations of on-board sensors to provide information to allow for the CAV to understand its environment and make path planning decisions accordingly mean that connectivity can be seen as a complementary paradigm. The three main areas where connectivity can complement on-board sensors and enable CAVs, discussed further in the following sections, are (1) crash avoidance, (2) mobility enhancement, and (3) environmental impact.

Motivations: Crash Avoidance

The primary motivation for equipping vehicles with connectivity technology is to enhance public safety through crash avoidance. There are some 5 million crashes, 35k fatalities, and 2.4M injuries annually in the United States (National Highway Traffic Safety Administration, 2018). In order to reduce the number and severity of collisions, the automotive industry has implemented passive safety technologies like seat belts and airbags, as well as active safety technologies such as ESC and AEB. The active safety technologies have mostly helped vehicles take measures to avoid an accident whereas connectivity promises to help drivers react to collision-imminent situations. It does so by providing information that would not otherwise be available via sensors and/or drivers' own vision, i.e., connectivity can provide additional information not otherwise available because "line-of-sight" is not necessary. Connectivity can also provide a level of redundancy if various entities in the CAV's environment provide information on their position, velocity, and acceleration.

There are numerous scenarios in which connectivity can provide additional and/or redundant information. Connectivity can provide safety information to a single vehicle to avoid a crash, such as a speed advisory in the case of an accident or dangerous situation (i.e., an icy road) upstream from the vehicle, as shown in Figure 3.4 for a vehicle approaching a curve in the road.

FIGURE 3.4 Speed advisory depiction.

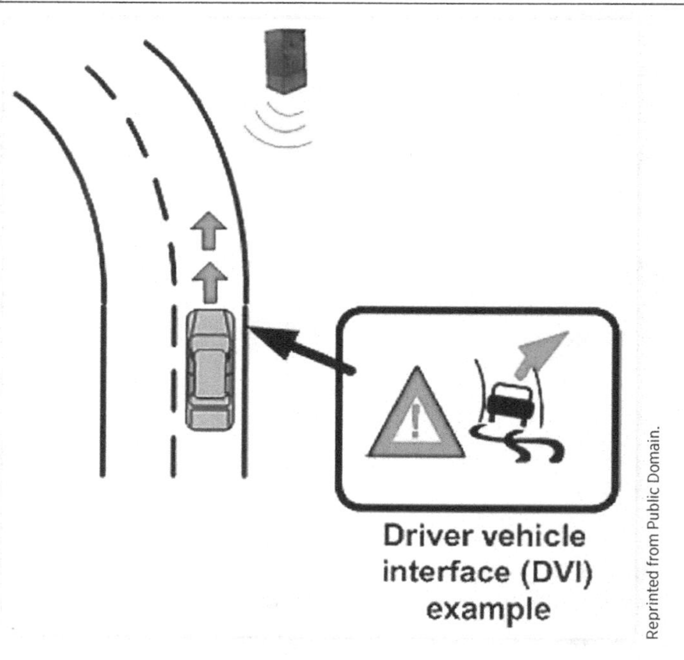

Example applications involving multiple vehicles include:

- Intersection Movement Assist (IMA): A CAV approaching an intersection is warned if another vehicle in or near the intersection is making a sudden turn or "running" a red light. The communication could happen via an intersection management system (V2I) or directly between vehicles (V2V).
- Left Turn Assist (LTA): A CAV attempting to make a left turn at an intersection is warned if the turn is likely to coincide with oncoming traffic to either abort the turn or complete the turn more quickly than originally planned. Similarly, the communication could happen via an intersection management system (V2I) or directly between vehicles (V2V), although the former is more likely in this case.

The NHTSA estimates that just IMA and LTA could prevent 592,000 crashes per year and save 1,083 lives (Laws, 2014). It's clear that connectivity could significantly reduce the chronic problem of vehicular injuries and fatalities.

V2X can also reduce conflicts between different types of traffic participants, including VRUs such as pedestrians, cyclists, and scooter operators. The World Health Organization estimates that VRUs account for approximately half of the 1.2M yearly fatalities around the world, with children and the elderly overrepresented in the statistic (World Health Organization, 2009).

Another use case for connectivity with respect to crash avoidance is teleoperation. In this scenario, a remote driver could take control of the vehicle if the driver or ADS is incapacitated or otherwise unable to complete the DDT. In 2021, Waymo has an NOC with remote drivers monitoring the CAVs deployed in the robo-taxi service in the Phoenix, AZ, metropolitan area, ready to take over the DDT to avoid accidents and increase the overall safety of the service.

The case for connectivity can largely be made on crash avoidance alone. V2I technology can reduce single-vehicle accidents: the NHTSA has identified 15 of this type, responsible for approximately 27% of accidents. V2V technology can reduce multiple-vehicle accidents: the NHTSA has identified 22 of this type, responsible for some 81% of accidents. Connectivity introduces a paradigm shift in automotive safety from passive (or reactive) safety measures (i.e., airbags, seatbelts) to active (or pro-active) safety measures that can reduce crashes and all of the negative impacts to society that crashes bring.

Motivations: Mobility Enhancement

Another main motivation for equipping vehicles with connectivity technology is to help increase mobility in a traffic network, in part by alleviating traffic and the associated time wasted in traffic. The scope of the traffic problem can be seen in a Texas A&M Transportation Institute estimation that in 2017, the average U.S. commuter spent 54 hours in traffic (up from 34 hours in 2010); this is equivalent to 1.4 full work (or vacation) weeks for each driver (Schrank, Eisele, & Lomax, 2019).

Connectivity can assist in mobility enhancement and traffic reduction in myriad ways, many of them seemingly small. Some example applications include:

- Route guidance (V2I): CAVs could receive information on traffic flows, thereby allowing the CAV to choose an optimal path. This is an improved version of what human drivers using navigation applications currently do; however, CAVs can examine more information and make better evidence-based decisions more quickly. Further, if there is a central traffic management system that is communicating with vehicles on the road, with enough participation, the entire traffic network flow could be optimized.

- Queue warning (V2I): More targeted than route guidance, in this case, the CAV receives information on a traffic buildup developing in a lane, so the CAV can decrease speed, change lanes, or divert to a different route.

- Connection protection (V2I): While not necessarily CAV related, this application allows someone planning to take public transit to request a "hold" of a transit vehicle via connectivity so that the connection can be made. This increases public transit usage, which, in turn, increases mobility.

- Disabled pedestrian (V2P): Also not necessarily CAV related, this application allows for disabled pedestrians to communicate their intentions to cross a street via V2P connectivity (perhaps with an intersection management system that communicates with CAVs or with the CAVs directly) that enables safer movement for such pedestrians, thereby enhancing mobility.

- Platooning (V2V): Connectivity between CAVs could allow for shorter gaps left between vehicles and fewer brake events that cause so-called traffic snakes leading to congestion. Platooning, depicted in an "eco-lane" in Figure 3.5, can increase the throughput of a traffic network without adding any additional roads.

FIGURE 3.5 Platooning of connected vehicles in an "eco-lane."

Reprinted from Public Domain.

The U.S. DOT estimates that connectivity could reduce traffic time delays due to congestion by one-third, which could significantly improve mobility in the traffic network.

Motivations: Environmental Impact

The third main motivation for connectivity in vehicles is to reduce the environmental impact of the transportation system. Connectivity can assist in the reduction effort by reducing the time spent idling (while EVs do not pollute when not in motion, conventional ICE-based vehicles pollute when stopped and the engine is idling, although start-stop systems have reduced this problem somewhat). The Texas A&M Transportation Institute estimated that in 2017, 3.3 billion gallons of fuel were wasted in traffic (up from 1.9 billion gallons in 2010), meaning that the resulting emissions from those vehicles did not serve any mobility purpose. Moreover, the cost of the time in terms of lost productivity and the cost of fuel together total over $179B per year, or $1,080 per commuter (up from $700 in 2010) (Schrank, Eisele, & Lomax, 2019). The impact of idling vehicles, vividly shown in Figure 3.6, or those in stop-and-go traffic releases more GHGs and CACs per mile traveled than vehicles moving at 30-45 mph (Cortright, 2017).

FIGURE 3.6 Idling vehicles.

NadyGinzburg/Shutterstock.com.

Connectivity allows travelers to make more informed travel choices to minimize environmental impact as well as impact the interactions with the traffic network infrastructure:

- Travelers could choose alternate routes, modify departure times, or even modify the travel method.
- Traffic signals could be adjusted in real time to maximize traffic flow and reduce stops/starts and acceleration, and even suggest/demand that CAVs use a specific approach speed (i.e., in a CDA situation) for the intersection while providing SPaT information, as shown in Figure 3.7. The better the coordination between vehicles the higher the overall energy efficiency of the vehicles (as discussed below in the Connectivity Case Study: ACC versus CACC section). However, the modulated speeds of the connected vehicles can also impact the surrounding vehicles positively to reduce braking events and improve traffic flow and overall traffic efficiency, which reduces vehicle emissions and energy consumption. This would also have the side effect of improving public transit reliability, thereby potentially creating a virtuous cycle where more people use public transit and reduce the number of vehicles on the road.
- Priority could be enabled for transit vehicles, allowing for faster movement through the network, and thereby increasing the appeal of public transit.
- Differentiation of environmental impact between vehicles (i.e., EVs versus conventional, ICEVs) and allow for dynamic "Eco-lanes" to be created on highways, thereby improving upon existing HOV lane environmental impact through connectivity to identify high-occupancy, low-impact vehicles.

FIGURE 3.7 SPaT information connectivity depiction.

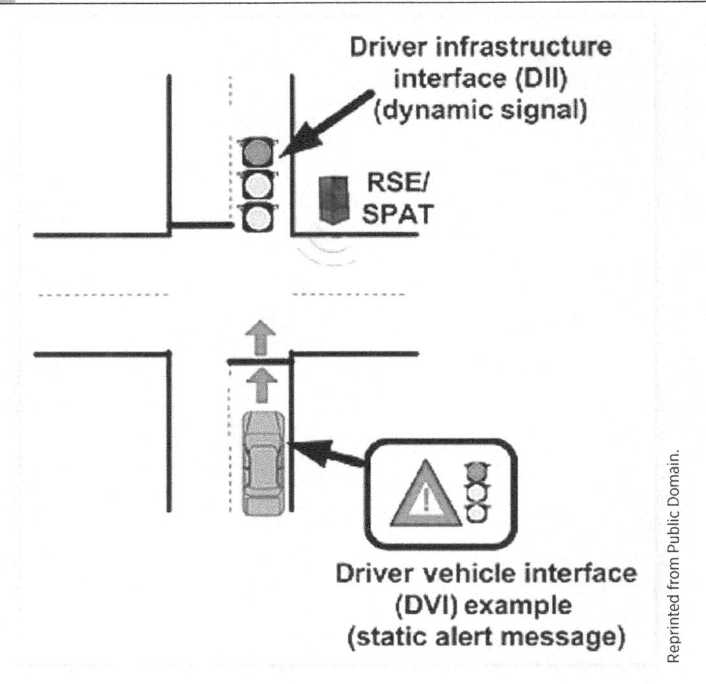

Connectivity Case Study: ACC versus CACC

One of the use cases for connectivity is to enhance the advantage of using adaptive cruise control (ACC) systems to modulate the flow of traffic and reduce brake events that cause a phenomenon known as a "traffic snake," in which the wave of braking propagates downstream in traffic. If the wave is sufficiently severe, a traffic jam occurs. ACC systems reduce traffic snakes by reducing acceleration and braking rates and maintaining the distance to the vehicle ahead. The usage of an ACC system can also improve the energy efficiency of a vehicle through a reduction in speed variation and of acceleration rates. Adding connectivity creates what is known as a cooperative adaptive cruise control (CACC) system. A CACC system allows for even greater speed variation and acceleration rate reductions due to vehicles further upstream relaying braking and throttle information to the vehicles downstream, which allows the latter vehicles to plan accordingly and maximize efficiency.

The improvement of CACC versus ACC is demonstrated in Figure 3.8. This figure (Milanes, et al., 2014) depicts a four-car platoon in which the lead car followed a series of speed changes meant to emulate real-world traffic:

1. Start with a speed of 25.5 m/s.
2. Accelerate at a rate of (1/80)g until 29.5 m/s, remain at 29.5 m/s for 10 s.
3. Decelerate at a rate of (1/80)g until 25.5 m/s, remain at 25.5 m/s for 10 s.
4. Accelerate at a rate of (1/40)g until 29.5 m/s, remain at 29.5 m/s for 15 s.
5. Decelerate at a rate of (1/40)g until 25.5 m/s, remain at 25.5 m/s for 15 s.

6. Accelerate at a rate of (1/20)g until 29.5 m/s, remain at 29.5 m/s for 20 s.
7. Decelerate at a rate of (1/20)g until 25.5 m/s, remain at 25.5 m/s for 20 s.
8. Accelerate at a rate of (1/10)g until 29.5 m/s, remain at 29.5 m/s for 20 s.
9. Decelerate at a rate of (1/10)g until 25.5 m/s, remain at 25.5 m/s for 20 s.

The gap setting for the following vehicles in the platoon was 1.1 s (the shortest setting available for the test vehicles). As can be seen from Figure 3.8(a), the ACC systems were not able to maintain a stable gap between the platoon vehicles, with unequal speeds and accelerations of the vehicles throughout. When CACC is incorporated, as shown in Figure 3.8(b), the gaps (shorter than in the ACC case at 0.6 s) are maintained to a much higher degree, the vehicles have much more uniform speeds and acceleration rates (i.e., speed harmonization), and the efficiencies of the following vehicles and the overall platoon is much higher.

FIGURE 3.8 Speed, acceleration, and time graphs of a (a) four-vehicle platoon with ACC and (b) four-vehicle platoon with CACC.

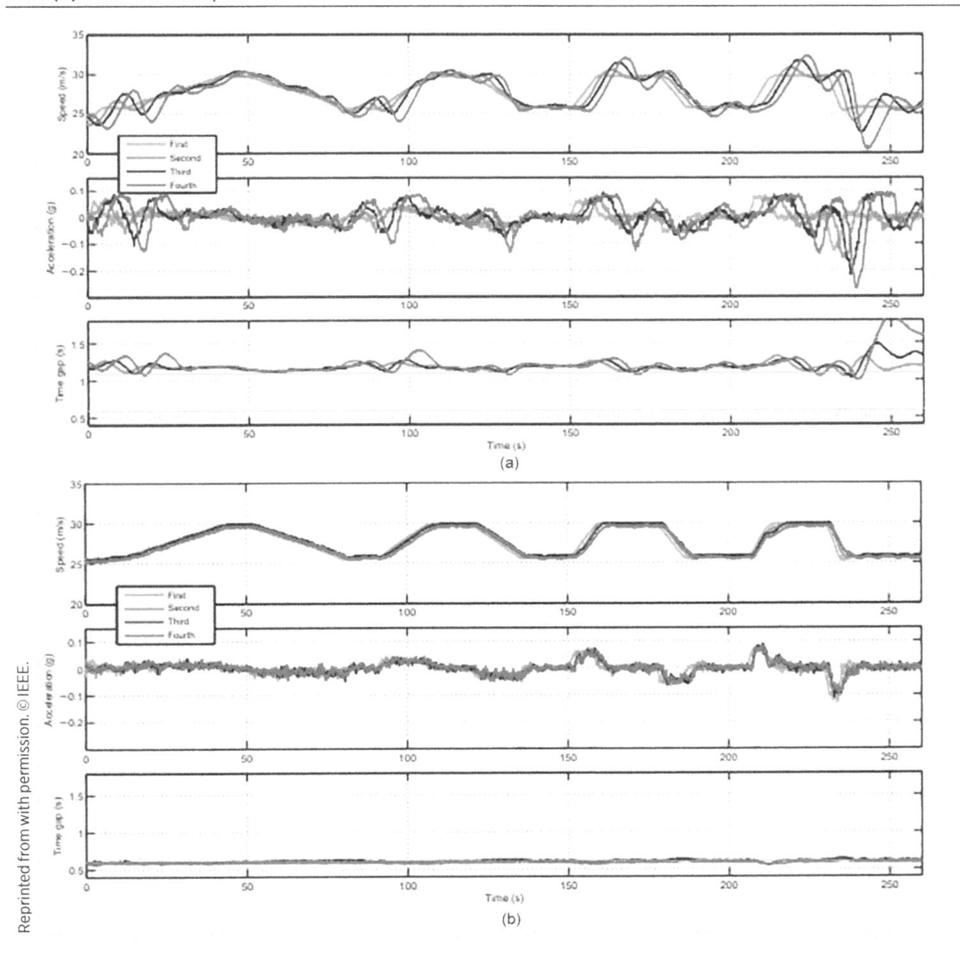

The efficiency improvements in the CACC platoon case due to a reduction in aerodynamic drag are not explored in Milanes, et al., 2014, but such improvements are quite possible. One study conducted by the NREL, Peloton, Intertek, and Link Engineering gives an indication of the possible aerodynamic improvement. The study consisted of one "control truck," one "lead truck," and one "following truck," (Lammert et al., 2014) shown in Figure 3.9. All were of the same make, model, and EPA SmartWay-compliant aerodynamic package. The control truck was driven on the opposite end of an oval track in Uvalde, TX, to the platooning trucks. The parameters of the platoon were then varied, including speed (55 mph to 70 mph), following distance (20 ft to 75 ft), and gross vehicle weights (GVWs; 65,000 lb to 80,000 lb). The maximum increases in fuel economy (the results were not for the same configuration of speed, following distance, and GVW) were:

- Lead vehicle 5.3%
- Following vehicle 9.7%
- Platoon 6.4%[1]

FIGURE 3.9 Platooning study trucks.

[1] Interestingly, the maximum platoon efficiency did not occur at the shortest following distance. This was due to the engine fan of the following vehicle drawing more power for engine cooling at that distance, resulting in diminishing returns. If the vehicles were EVs, it is possible that the shortest following distance would result in the greatest platoon efficiency due to minimization of aerodynamic drag.

It is clear that CACC is useful from an efficiency perspective. However, CACC is also useful from a safety perspective since the conveyance of impending brake actuation to following vehicles provides more time for these following vehicles to respond by initiating preemptive braking to avoid hard decelerations. It is important to note that platooning is not currently allowed in all jurisdictions, although this state of affairs is rapidly changing, as shown in Figure 3.10. In the figure, states in red have not yet (as of 2019) explicitly authorized platooning vehicles via a Federal Trade Commission rule exemption; states in yellow have taken limited steps to authorize platooning vehicles while maintaining some restrictions; states in green have fully authorized platooning vehicles without restrictions.

FIGURE 3.10 Platooning allowance by U.S. states.

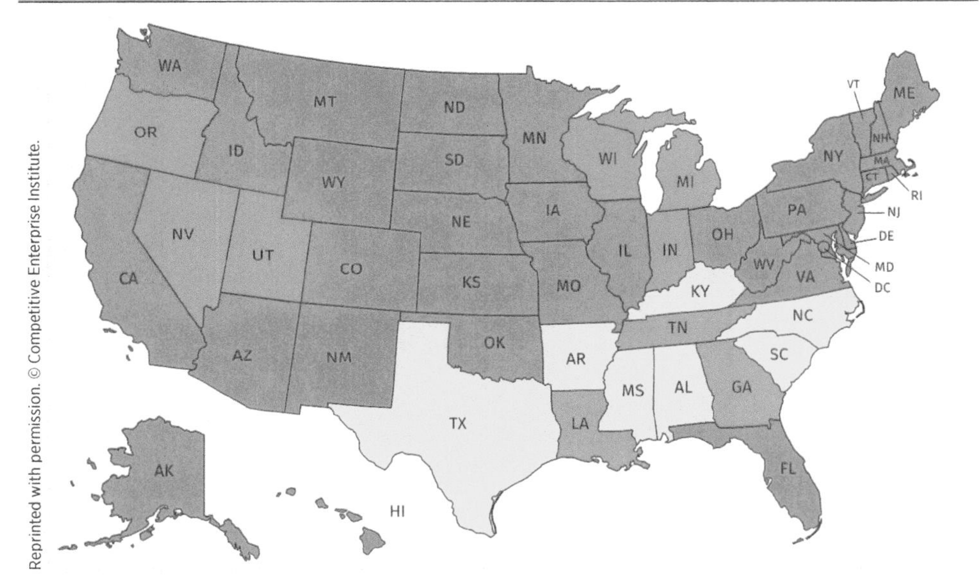

Connectivity Technology

There are several components of a V2X system regardless of the wireless protocol being used that include equipment both on-board and off-board the CAV. The U.S. DOT's Intelligent Transportation Systems office published an ePrimer that includes the following list (Hill & Krueger, 2012):

1. On-board equipment (OBE): The OBE or mobile equipment represent the systems or devices through which most end users will interact with the connected vehicle environment in order to gain the benefits of the anticipated safety, mobility, and environmental applications. The OBE includes the transceiver that enables V2X applications. In addition, other technologies associated with vehicles or mobile devices participating in the connected vehicle environment are necessary to provide basic information used in the various connected vehicle applications. This information includes vehicle or device location, speed, and heading that is derived from GPS or other sensors. Additional data from other vehicle sensors, such as

windshield wiper status, anti-lock braking system activation, or traction control system activation, may be beneficial in certain connected vehicle applications.

2. Roadside unit (RSU): This equipment will support three main types of functionality. First, it connects vehicles and roadside systems, such as systems integrated with traffic signal controllers, which allow users to participate in local applications such as intersection collision avoidance. Second, the RSU provides the connectivity between vehicles and network resources that are necessary to implement remote applications—for example, for supporting the collection of probe vehicle data used in traveler information applications. Third, the RSU may be required to support connected vehicle security management.
3. Core systems: These are the systems that enable the data exchange required to provide the set of connected vehicle applications with which various system users will interact. The core systems exist to facilitate interactions between vehicles, field infrastructure, and back-office users. Current thinking envisions a situation of locally and regionally oriented deployments that follow national standards to ensure that the essential capabilities are compatible no matter where the deployments are established.
4. Support systems: These include the security credential management systems (SCMSs) that allow devices and systems in the connected vehicle environment to establish trust relationships. Considerable research is underway to describe both the technical details and the policy and business issues associated with creating and operating a security credentials management system.
5. Communications systems: These comprise the data communications infrastructure needed to provide connectivity in the connected vehicle environment. This will include V2V and V2I connectivity and network connectivity from RSUs to other system components. These system components will include core systems, support systems, and application-specific systems. The communications systems will include the appropriate firewalls and other systems intended to protect the security and integrity of data transmission.
6. Application-specific systems: This refers to the equipment needed to support specific connected vehicle applications that are deployed at a particular location, rather than the core systems that facilitate overall data exchange within the connected vehicle environment. An example could be software systems and servers that acquire data from connected vehicles, generate travel times from that data, and integrate those travel times into TMC systems. Existing traffic management systems and other ITS assets can also form part of an overall connected vehicle application.

There are also communications latency requirements that are applicable regardless of the wireless protocol used. In general, latency requirements are more stringent for safety-critical applications like completion of the DDT by the CAV. Figure 3.11 shows how several wireless protocols (aka communications technologies) compare in terms of latency ranges. In 2021, there are competing wireless protocols in the V2X industry: The notable protocols in the V2X context are DSRC at 0.0002 s latency and Cellular with a range of 1.5-3.5 s of latency (more on the latter in the cellular V2X (C-V2X) section below). The figure also shows two thresholds for Active Safety (aka, colloquially, as Advanced Driver Assistance System (ADAS)) usage of between 0.02 s and 1 s. The DSRC meets the most stringent latency requirement while the Cellular technology does not meet even the least stringent latency requirement. The two protocols are discussed in more detail, including a head-to-head comparison, in the following sections.

FIGURE 3.11 Communications latency requirements for CAVs.

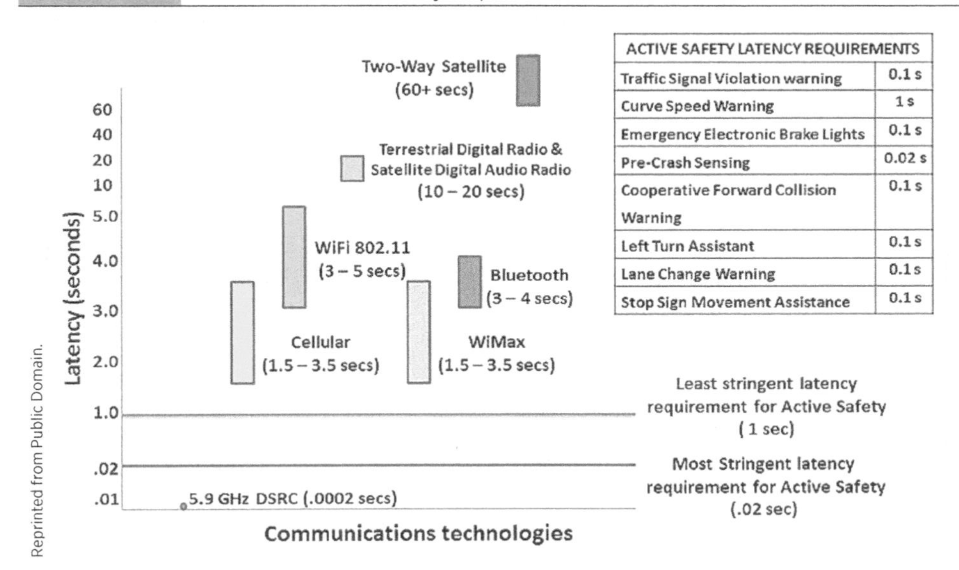

ACTIVE SAFETY LATENCY REQUIREMENTS	
Traffic Signal Violation warning	0.1 s
Curve Speed Warning	1 s
Emergency Electronic Brake Lights	0.1 s
Pre-Crash Sensing	0.02 s
Cooperative Forward Collision Warning	0.1 s
Left Turn Assistant	0.1 s
Lane Change Warning	0.1 s
Stop Sign Movement Assistance	0.1 s

Connectivity Technology: DSRC

In 2021, DSRC is the most widely deployed connectivity technology. V2X networks that use DSRC technology involve wireless transmission/reception (i.e., bidirectional) of messages within a 75 MHz section in the 5.9 GHz band of the radio frequency spectrum. This spectrum was established in 1999 by the U.S. FCC specifically for "intelligent transportation systems" (ITS). The European Telecommunications Standards Institute allocated a 30 MHz section in the same 5.9 GHz band for DSRC in 2008.

DSRC is an open-sourced protocol based on the WLAN IEEE 802.11 standards known as Wireless Access in Vehicular Environments (WAVE) in the United States that was completed in 2009. In Europe, the protocol is known as ITS-G5. The WAVE standard instructs the vehicle to issue information about its position and path information at a frequency of 10 Hz (Li, 2012). DSRC has several positive attributes:

- DSRC was designed for vehicular applications like V2X.
- DSRC has low latency and high reliability, i.e., short time delay and assurance of message delivery.
- DSRC is capable of being used to communicate at distances up to 1,000 ft.
- DSRC is robust to radio interference.

An example OBU and RSU DSRC pair is shown in Figure 3.12.

FIGURE 3.12 Example DSRC RSU and OBU devices.

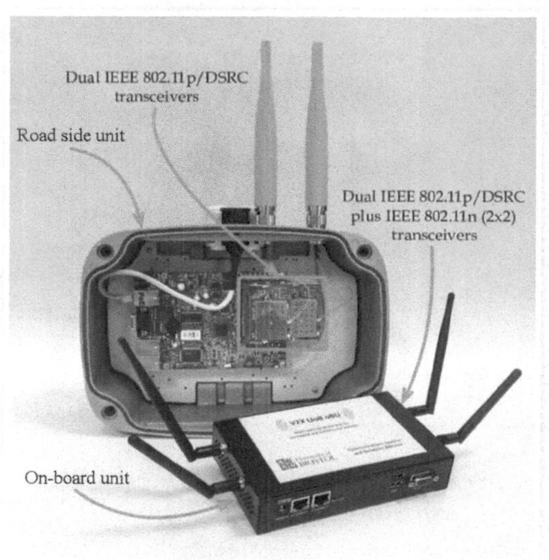

Connectivity Technology: C-V2X

The second competing connectivity technology is C-V2X, with 4G LTE cellular technology currently dominant and 5G beginning to be deployed. Worldwide deployment of 5G is depicted in Figure 3.13. Deployment was started in earnest by all four major U.S. cellular networks in 2019. 5G is the fifth-generation wireless technology, and the standard was agreed to by the wireless industry in December 2017. Compared to 4G LTE, 5G is designed to have greater data transfer speed, lower latency, and the ability to connect more devices at the same time. Since 4G LTE does not meet the latency requirements shown in Figure 3.11, the objective for 5G is to achieve a latency of 1 ms (along with a speed of 20 Gpbs).

FIGURE 3.13 Depiction of worldwide 5G deployment.

The advantages of C-V2X technology include longer range (initial tests have shown a range improvement of 20-30% over DSRC [Gettman, 2020]) and the usage of existing infrastructure rather than RSUs used only for CAVs as in the case of DSRC.

Connectivity Technology: DSRC versus 5G

The competition between DSRC and C-V2X technologies is analogous to the "VHS versus Betamax" situation in the 1980s or more recent "Blu-ray versus HD DVD" situation in the 2000s. These situations all involve competing technologies that serve essentially the same purpose, but with different technological approaches even if the overall specifications are similar. The resolution (famously in the case of VHS vs. Betamax) is not necessarily in favor of the superior technology, as there are many factors, including cost and user experience, that determine the winner.

From a specification standpoint, DSRC can accommodate all necessary V2X communications in commercially available technology. Regulators in Europe and the United States have also embraced DSRC. In fact, in 2016, the U.S. DOT proposed requiring all vehicles to be equipped with DSRC by 2023. Further, starting in 2017, vehicles and roads have already been equipped with DSRC in the United States. In 2019, the Volkswagen (VW) Golf was released with DSRC capability in Europe, the first mass-produced vehicle with this capability (Autotalks.com, n.d.).

While DSRC has the advantage of being the "first mover" in V2X technology, it does have some technological drawbacks, for example, it uses the same frequency band used by some Wi-Fi devices, and interference issues may arise. Further, in December 2020, the U.S. FCC rescinded the lower 45 MHz of the DSRC section of the 5.9 GHz band for unlicensed Wi-Fi applications and declared that the remaining 30 MHz of the spectrum would be allocated for C-V2X, and not DSRC. The controversial decision was justified, in part, due to the FCC's position that C-V2X can be used as an alternative for V2X applications but could be reversed by future U.S. federal administrations.[2]

C-V2X is the relative newcomer technology and has momentum with regulators (as seen by the FCC decision) and even vehicle manufacturers such as VW, but also the GSMA mobile phone consortium. China has also been strongly advocating for C-V2X over DSRC, with heavyweights Baidu and Huawei supporting the 5G route. C-V2X has some structural advantages, especially when 5G is the technology of choice. For instance, 5G has wider bandwidth and uses private cell phone towers rather than government equipment for RSUs. Additionally, 5G will be installed in vehicles for mobile entertainment, and in mobile phones for V2P possibilities. 5G's longer range means earlier alerts and better visibility of unexpected and potentially dangerous situations, allowing travel at higher speeds, encompassing broader ODDs. However, there is still the outstanding issue of latency as previously mentioned, and technological development work is needed to ensure that C-V2X systems can be trusted in safety-critical situations.

One possible result in this technology competition is a complementary system rather than a competitive one. In a bifurcated system, DSRC might handle the tasks requiring low latency, and 5G C-V2X might handle the longer-range tasks such as traffic congestion and HD map download. It is impossible to say at this point, although the 2020 FCC decision

---

[2] One of the FCC commissioners, Michael O'Reilly, noted in his decision that even though DSRC had been allocated the spectrum portion in 1999, that over 20 years later, only 15,000 vehicles were equipped with DSRC technology (with only 3,000 vehicles actually sold), meaning that the DSRC deployments have rarely been commercialized, and have been mostly experimental) (Fiercewireless.com, 2020).

has made a vehicle connectivity future of C-V2X more likely. If there is a winning technology, a negative situation of stranded assets (after significant capital expenditure) could result.

Connectivity Technology: CV Costs

One of the major issues with connectivity is equipment cost, and who will be paying for this equipment. For the OBUs, the owner of the CAV will pay for the equipment, for example, either upfront in the purchase price or as part of the lease. The source of funding for the RSUs and supporting equipment is less clear. It is unclear who will provide the funding and whether the funding will be at sufficient scale and timeliness in order for connectivity to be impactful in the transportation sector. It remains an open question whether the RSUs and supporting equipment will be funded by private or public entities, and how these technologies will be owned, operated, and regulated.

The costs will also depend on the particular technology. In 2016, the NHTSA estimated that V2V technology (i.e., DSRC) and supporting communications functions (including a security management system) would cost approximately $350 per vehicle in 2020, decreasing to $227 by 2058 (US Department of Transportation—National Highway Traffic Safety Administration, 2016). This cost estimate includes the increase in expected fuel costs from the V2V equipment weight. The security management system was expected to cost $6 per vehicle, which could increase over time as more vehicles are included in the V2V system. Noticeably absent were any costs associated with data storage and transfer as these costs are not free. The RSU costs must be considered as well and may place strain on government budgets with an estimated price tag of $50k per unit.

Deployment Challenges versus Potential Benefits

As mentioned at the beginning of this chapter, connectivity (beyond features using cellular for navigation and internet access) has been deployed more slowly than anticipated or desired by many in the automotive industry. This is due to the myriad of challenges and obstacles that have yet to be completely addressed. The U.S. Government Accountability Office (GAO) identified the following V2I deployment challenges in a 2015 report (quoted here [GAO, 2015]):

1. Ensuring that possible sharing with other wireless users of the radio-frequency spectrum used by V2I communications will not adversely affect V2I technologies' performance.
2. Addressing states and local agencies' lack of resources to deploy and maintain V2I technologies.
3. Developing technical standards to ensure interoperability.
4. Developing and managing data security and addressing public perceptions related to privacy.
5. Ensuring that drivers respond appropriately to V2I warnings.
6. Addressing the uncertainties related to potential liability issues posed by V2I.

General V2X deployment challenges (in addition to those listed by the GAO above) include:

1. Developing the SCMS to avoid hacking.
2. Addressing the costs of V2X systems, especially for low-income vehicle owners.
3. The cost of data is not negligible and will grow as more information is shared. The amount of data has been estimated at 25 GB for a single CAV per hour, and 130 TB per year (SmartCitiesWorld Webinar, 2021).
4. Addressing public acceptance issues.
5. Ensuring that the connectivity is consistently reliable and that the latency issue discussed earlier is addressed.
6. Having both connectivity and automation features on board a vehicle increases the data processing requirements since the data from both streams must be synchronized and merged. However, this can improve redundancy of the perception system and, therefore, increase safety.
7. Securing a unique cooperation between disparate entities with varying connections and mandates:
    a. Federal, state, and local transportation authorities
    b. OEMs (cars, buses, trucks)
    c. Telecommunications providers
    d. Electronics manufacturers
    e. Academia
    f. Public

Despite this somewhat daunting list of challenges that are largely outstanding and in need of addressing, there is considerable interest in connectivity in the automotive industry due to the potential benefits that are seen as manifesting ever faster and with deeper impact as more vehicles with connectivity are deployed. In part, the enthusiasm stems from the benefits listed above under the categories of crash avoidance, mobility enhancement, and environmental impact, which do not capture all of the potential benefits. Further possible benefits comprise a wide and diverse spectrum:

- Connectivity could eventually allow for the elimination of street signs and traffic lights, with edge-computing intersection managers directing traffic for reduced infrastructure costs.
- EV owners searching for EVSE could know which locations are closest to their route or destination and if the EVSE are in use. Reservations could even be made to ensure that the EVSE is available upon arrival.
- Fleet managers can always keep track of their vehicles, including diagnostics and driving behavior of their drivers. This monitoring could improve the safety and efficiency of fleet operations.
- OEMs can introduce new features and update the software in their vehicles. Pioneered by Tesla, over-the-air (OTA) updates have allowed the company to extend the time between model generations beyond the normal five-year (or so) cycle of other OEMs. Without having to visit a dealer, Tesla owners receive new (and sometimes surprising) features to keep the vehicle experience novel.

This interest means that despite low commercial deployment, companies continue to invest and partner to share costs and expertise as the development of connectivity technology expands. There is significant interconnectedness in the connected vehicle industry. Many companies have a significant number of partners, often overlapping with competitors.

The future of connectivity, while not assured, is seen by many in the automotive industry to be promising although the clear industry leaders are yet to emerge, as is the winning technology. The next decade is critical for this industry to consolidate and agree on technology in addition to generating uniformity throughout the industry as deployments occur.

References

Autotalks.com. (n.d.). DSRC vs. C-V2X for Safety Applications.

Cadillac Customer Experience. (2017, March 9). V2V Safety Technology Now Standard on Cadillac CTS Sedans. Retrieved from Cadillac Pressroom: https://media.cadillac.com/media/us/en/cadillac/news.detail.html/content/Pages/news/us/en/2017/mar/0309-v2v.html

Cortright, J. (2017, July 6). Urban Myth Busting: Congestion, Idling, and Carbon Emissions. Retrieved from Streetsblog USA: https://usa.streetsblog.org/2017/07/06/urban-myth-busting-congestion-idling-and-carbon-emissions

Fiercewireless.com. (2020, November 18). FCC Votes to Open 5.9 GHz for Wi-Fi, C-V2X.

Gettman, D. (2020, June 3). DSRC and C-V2X: Similarities, Differences, and the Future of Connected Vehicles. Retrieved from Kimley-Horn.com: https://www.kimley-horn.com/dsrc-cv2x-comparison-future-connected-vehicles

Government Accountability Office. (2015). "Intelligent Transportation Systems: Vehicle-to-Infrastructure Technologies Expected to Offer Benefits, but Deployment Challenges Exist," Report to Congressional Requesters.

Hill, C., & Krueger, G. (2012). ITS ePrimer—Module 13: Connected Vehicles. U.S. Department of Transportation—Intelligent Transportation Systems.

Howard, B. (2014, March 13). Audi A3 Is First Car with Embedded 4G LTE—But Will Owners Go Broke Streaming Movies? Retrieved 12 4, 2020, from Extremetech: https://www.extremetech.com/extreme/178416-audi-a3-is-first-car-with-embedded-4g-lte-but-will-owners-go-broke-streaming-movies.

Jaillet, J. (2013, July 26). NTSB Asks for 'Connected Vehicles' Mandate. Retrieved from Commercial Carrier Journal: https://www.ccjdigital.com/business/article/14927741/ntsb-asks-for-connected-vehicles-mandate

Lammert, M., Duran, A., Diez, J., Burton, K., & Nicholson, A. (2014). Effect of Platooning on Fuel Consumption of Class 8 Vehicles Over a Range of Speeds, Following Distances, and Mass. *SAE Commercial Vehicle Congress*. Rosemont, IL, Paper 2014-01-2438.

Laws, J. (2014, November 1). Revving up V2V. Retrieved from Occupational Health & Safety Online: https://ohsonline.com/Articles/2014/11/01/Revving-Up-V2V.aspx

Li, Y. (2012). An Overview of DSRC/WAVE Technology. In Zhang, X., & Qiao, D., *Quality, Reliability, Security and Robustness in Heterogeneous Networks*. Springer, Berlin.

Maricopa County Department of Transportation. (2017). Connected Vehicle Test Bed: Summary.

Milanes, V., Shladover, S., Spring, J., Nowakowski, C., Kawazoe, H., & Nakamura, M. (2014). Cooperative Adaptive Cruise Control in Real Traffic. *IEEE Transactions on Intelligent Transportation Systems*, 15, 296-305. doi:10.1109/TITS.2013.2278494

National Highway Traffic Safety Administration. (2018). DOT HS 812 451—Quick Facts 2016.

SAE International. (2020, July). J2735_202007—V2X Communications Message Set Dictionary.

SAE International Surface Vehicle Information Report. (2020, May). J3216_202005—Taxonomy and Definitions for Terms Related to Cooperative Driving Automation for On-Road Motor Vehicles.

Schrank, D., Eisele, B., & Lomax, T. (2019). Urban Mobility Report. Texas Transportation Institute.

Shepardson, D. (2019, April 26). Toyota Halts Plan to Install U.S Connected Vehicle Tech by 2021. Retrieved from Reuters: https://www.reuters.com/article/autos-toyota-communication/toyota-halts-plan-to-install-u-s-connected-vehicle-tech-by-2021-idUSL1N22816B

Slovick, M. (2018, May 26). Toyota, Lexus Commit to DSRC V2X Starting in 2021. Retrieved from Innovation Destination—Automotive: https://innovation-destination.com/2018/05/16/toyota-lexus-commit-to-dsrc-v2x-starting-in-2021/

SmartCitiesWorld Webinar. (2021, June 9). Connected and Autonomous Vehicles: How to Deal with Data?

U.S. Department of Transportation. (n.d.). Vehicle-to-Infrastructure (V2I) Resources. Retrieved from Intelligent Transportation Systems—Joint Program Office: https://www.its.dot.gov/v2i/

US Department of Transportation—National Highway Traffic Safety Administration. (2016). 49 CFR Part 571—Federal Motor Vehicle Safety Standards, V2V Communications—Notice of Proposed Rulemaking (NPRM).

US Department of Transportation. (2016). Smart City Challenge Lessons Learned.

Walz, E. (2020, February 5). Audi Vehicles Can Now Communicate with Traffic Lights in Düsseldorf, Germany. Retrieved from FutureCar.com: https://www.futurecar.com/3766/Audi-Vehicles-Can-Now-Communicate-with-Traffic-Lights-in-Dsseldorf-Germany

White, A. (2020, December 15). The FCC Just Upended Decades of Research on Connected Vehicles. Retrieved from Car and Driver: https://www.caranddriver.com/news/a34963287/fcc-connected-cars-regulations-change-revealed/

World Health Organization. (2009). Global Status Report on Road Safety.

VOL
MAP
RADIO
MEDIA
NAV
FRONT
REAR
PASSENGER AIR BAG
A/C
CLIMATE
MODE
DUAL
CleanAir
TEMP
AUTO
TEMP

4

Sensor and Actuator Hardware

The breakthrough, application, and integration of various sensors on a CAV are key considerations that will make automated driving an integral part of our future transportation systems. Different sensors have distinct working principles, application areas, strengths, and weaknesses. Taking Figure 1.30 in Chapter 1 as an example, the red CAV will need different sensors with different FOVs and features to detect the pedestrian, the green CAV on the front side, the blue CAV on the right side, stop signs, and traffic lanes. As a key enabling technology of CAVs, familiarity with the capabilities of various sensors is necessary for a fundamental understanding of CAVs.

In this chapter, different sensors that are commonly used in CAVs will be introduced along with their characteristics, advantages, limitations, and applications. Vehicle actuation systems and powertrain components, which generally garner less attention than sensors when discussing CAVs, are also briefly described.

Principles and Characteristics of Sensor Hardware

Cameras

Definition and Description

Cameras are optical sensors and offer functionality analogous to the human eye. Their main purpose on CAVs is to mimic the human eyes' ability to visually sense the environment. A single camera operating alone, called a mono-vision system, results in an image that is 2D. Some CAV sensing requirements, such as the detections of lane boundaries, stop signs, and traffic lights, are well satisfied by abstracting key information from 2D images. However, because a mono-vision system cannot convey information in 3D, it cannot provide accurate information on how far away an object is from the camera, which is a functionality that is critical for safe CAV operation.

When two or more cameras are integrated, the result is a stereo-vision system. A 3D image can be obtained, and distance measurements become possible. This functionality is colloquially known as "depth perception." Handling 3D stereo-vision image information requires the use of more advanced and complex algorithms than the 2D mono-vision case, but can provide much of the required perception capability for CAV operation. Cameras are widely used on CAVs and may be considered as their most important sensor modality. Some companies, notably Tesla, have historically claimed that CAVs can be operated solely with camera-based sensing alone, and continue to develop ADS without other sensor modalities.

Most contemporary cameras, including those on CAVs, are digital cameras, meaning that image pickup is done electronically rather than chemically, and the photograph is stored in digital memory. Well-established image processing and/or AI-based computer vision technologies can be directly applied to process digital images. Although different cameras may capture images with varying image quality, most commercially available cameras can be used for CAV development and testing purposes. For example, Forward-Looking InfraRed (FLIR) cameras, shown in Figure 4.1, were used in a Formula-SAE CAV. Some cheaper cameras, like the Logitech C920 webcam, can also be applied for some initial development and tests. Figure 4.2, which shows a test vehicle developed at Arizona State University (ASU), displays that even these basic cameras can be successfully employed for functions such as lane boundary detection.

FIGURE 4.1 FLIR cameras used in a Formula-SAE CAV.

FIGURE 4.2 Logitech C920 Webcam was used for lane boundary detection in the development of a CAV at ASU.

Characteristics and Capabilities

Digital cameras, first commercialized in the late 1980s and 1990s, have since been well established for many applications that require surveillance or monitoring the environment. CAVs represent another natural use case for these devices. One main strength of cameras over other types of sensors that will be discussed in this chapter is their relatively low cost. In the context of CAV hardware, they range from cheap (multiple hundreds of dollars, e.g., Blackfly) to even cheaper (less than one hundred dollars, e.g., Logitech webcam).

Unlike active sensors such as a RADAR, a camera is a passive sensor and does not need to emit any signals (e.g., sound or light waves) to collect information. In the daytime, cameras typically have a long and clear detection range (over 100 m), which is superior to that of many other sensor types. However, they are limited by obstacles, FOV, sensor size, and focal length. Cameras can detect color information, which is a strength for CAV applications as some important objects in the environment, such as traffic lights, convey color-based information. Moreover, camera measurement can achieve high resolution with advanced sensors. Due to their color detection and resolution, cameras offer superior object detection capabilities than other types of sensors. They also offer improved classification of objects such as traffic control devices, obstacles, and other road users.

Although cameras have many strengths and advantages, some issues also limit their application in AVs. For one, camera measurement is severely affected by environment light levels or variation in ambient light. Shadows on surrounding objects and lane boundaries, sharp glare from the sun (when driving toward the sun's direction), and/or dark views at night will all introduce detection issues for cameras. While headlights can offer illumination to improve performance at night, some environmental conditions, such as rain, snow, and fog, can significantly impact the camera detection and may not be overcome by the use of headlights. Another issue related to camera use is the related software. Some advanced image processing and/or AI-based computer vision algorithms designed to handle the camera information require substantial processing time and power, which causes some challenges for the real-time implementation of perception algorithms and increases the overall energy consumption of CAVs.

A spider chart of camera characteristics, shown in Figure 4.3, summarizes cameras' strengths and weaknesses for CAV applications. In sum, the camera hardware is a simple yet critical sensor modality for CAVs, offering unique advantages such as low cost and color detection. Some of the limitations of cameras, such as detection in dark environments, could be improved or partially resolved through further algorithm and software development. An important advantage that must be mentioned as well is that camera images can be readily and easily interpreted by humans for CAV development, validation, and auditing purposes. They are thus likely to remain an integral part of CAV sensor suites in the foreseeable future.

FIGURE 4.3 Spider chart of a camera.

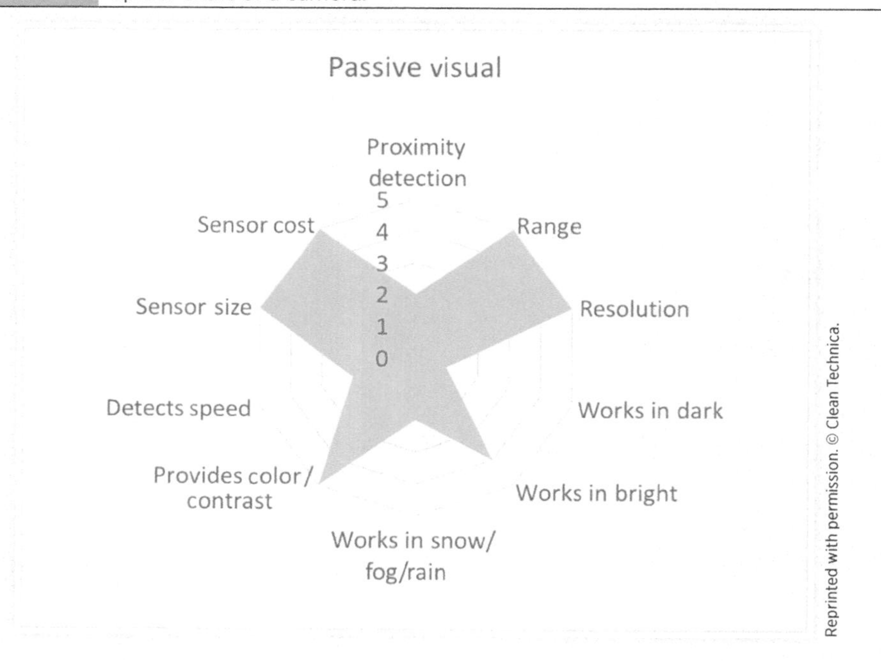

RADAR

Definition and Description

RADAR technology was developed independently by several nations before and during WWII for military applications. As implied by its full name, a RADAR is an object detection system that uses radio waves to determine the range and motion of an object. RADAR is one type of time-of-flight (TOF) sensor, which measures the time taken by an object, particle, or wave to travel a distance through a medium. Specifically, a RADAR utilizes the Doppler Effect to measure motions of detected objects by emitting and receiving radio waves, as shown in Figure 4.4. A typical RADAR system includes a transmitter, waveguide, duplexer (single antenna systems), receiver, and signal processor. An antenna transmits the

FIGURE 4.4 Emitted and reflected radio waves of a RADAR unit.

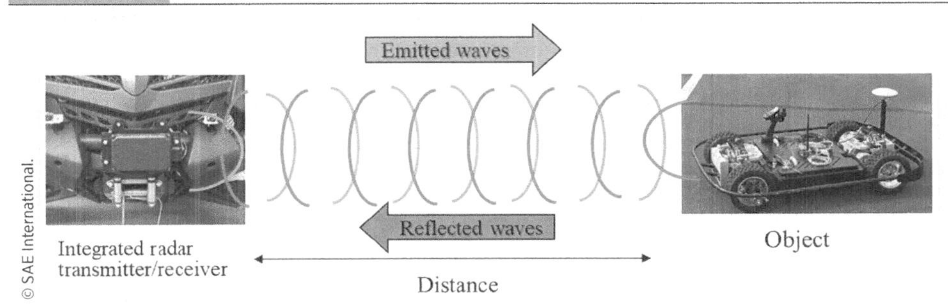

radio wave, which is reflected off the object and returned to the RADAR system. The reflected wave was captured and sent to the receiver, and then the signal is processed. The timing and characteristics of the reflected wave are used to determine the object's position and velocity. RADAR systems employ one of two main processing methods. The first is the direct propagation method, in which the delay associated with the reception of the reflected wave is measured for signal processing. The second is the indirect propagation method, also known as the frequency modulated continuous wave (FMCW) method. In FMCW, a modulated frequency wave is transmitted and received. Then, instead of time delay, the change in frequency is measured to determine the distances and relative speeds of objects. This approach tends to provide more accurate results. The FMCW uses different transmitting and receiving antennae to prevent the transmission signal from leaking into the receiver.

Characteristics and Capabilities

RADAR is used in a variety of fields, including aviation, shipping, meteorology, geology, traffic enforcement, office lighting, retail doors, and alarm systems. The RADAR system's size depends on the detection purpose and required detection range. Crucially, different applications employ different radio frequencies (or wavelengths). For sensing in automotive systems, the RADAR size is relatively small. Millimeter waves with a radio wavelength in the range of 1-10 mm are typically used. In terms of frequency, 24 GHz or 77 GHz are two frequencies most often used in automotive systems. A 24 GHz RADAR has a short range but a wide FOV of around 90° and is generally less expensive. They are typically used in the rear side of CAVs in multiple units. A 77 GHz RADAR has a long range but a narrow FOV (e.g., 15°), is more expensive, and is more often used in the front side of CAVs in a single unit.

Mass-produced RADAR units used in active safety system-equipped modern vehicles and CAVs cost a few hundred dollars. A typical RADAR can detect object motion up to 250 m away (higher functioning units can capture up to 1000 m). Considering this combination of performance and price, they are often considered the sensory type best suited for motion capture. The detection accuracy of (relative) position and speed of objects by RADAR is much higher than a camera. RADAR units can measure directly measure relative velocity, while for cameras, this is achieved by processing and comparing consecutive frames of images. RADAR units also work well in most driving environments, including snow, fog, and rain. They are influenced neither by sun and glare nor by low light conditions since they emit their own signals and do not rely on ambient illumination. Since the information most often required from RADAR units are positions and speeds of (multiple) objects, they can be computed relatively quickly without complex software processes. The energy or power consumption of RADAR units and associated processing is typically low, in the order of low, single-digit watts. Moreover, since a RADAR uses radio waves for detection with wavelengths in the millimeter range, some occluded objects may be detected and ranged physically by using the material-penetrating property of radio waves or even reflected waves from surfaces. Standalone RADAR units are widely used in today's vehicles for active safety system features such as AEB, FCW, and/or other crash avoidance systems, as well as in driving automation features like ACC. Figure 4.5 shows a Delphi ESR, installed in the front of a CAV, being successfully employed as part of an ACC system on public roads.

FIGURE 4.5 Delphi ESR was used in an ACC application.

Compared with cameras, the measurement resolutions of RADAR units are low. A common RADAR may not be able to construct a detailed 3D map or even a clear 2D image. Thus RADAR use cases are generally limited to the detection of objects and road users. Moreover, since RADAR is an active sensor, there is a risk of signal interference from other, nearby signal sources.

In sum, RADAR units have unique strengths to detect (relative) positions and speeds of moving objects, which is a key functionality of CAV perception systems. Some of the discussed limitations of RADAR units could be partially resolved through new hardware design and development, such as cascaded RADAR units to improve the detection resolutions. The spider chart of RADAR shown in Figure 4.6 summarizes several of the strengths and weaknesses discussed in this section.

FIGURE 4.6 Spider chart of a RADAR sensor.

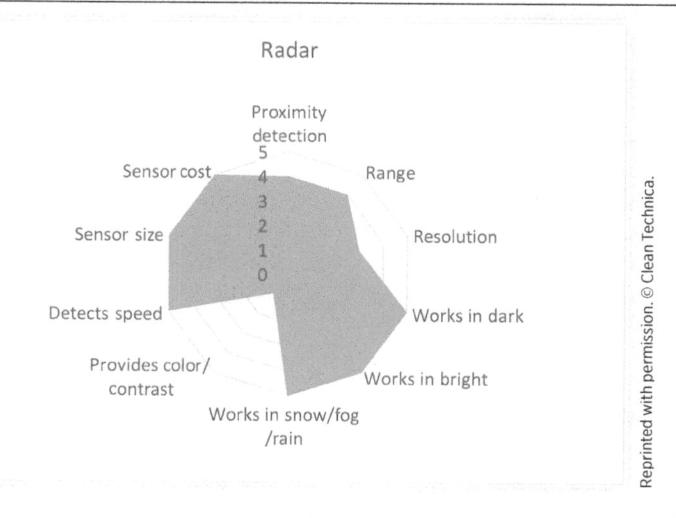

LIDAR

Definition and Description

Like "RADAR," "LIDAR" is an acronym. It stands for *Light Detection And Ranging.* Sometimes known as 3D laser scanning, LIDAR shares key similarities with RADAR, but uses a different part of the electromagnetic spectrum. LIDAR technology was first developed in the 1960s along with the advent of the LASER (light amplification by stimulated emission of radiation). Since then, LIDAR technologies have been widely applied in archeology, agriculture, and geology. Figure 4.7 shows an example of drone-mounted LIDAR units being employed to construct topographical maps.

FIGURE 4.7 Visualization of the VQ-840-G mounted on a drone in operation. The scan pattern on the ground is indicated by the green ellipse. The inserts show typical pulse shapes of the outgoing laser pulse and the echo return when measuring into the water.

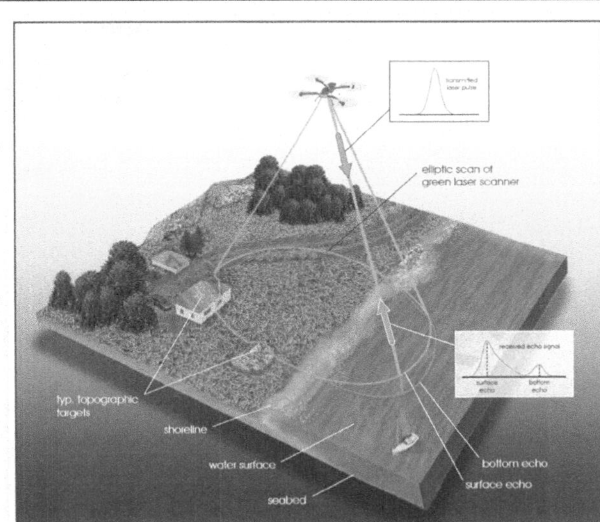

LIDAR was first introduced for CAV applications in the 2000s. In the 2005 Grand DARPA Challenge, a 2D LIDAR SICK was used in the Team Overbot, the CAV depicted in Figure 4.8. In the 2007 DARPA Challenge, 3D LIDAR units from Velodyne, like the one mounted atop the vehicle in Figure 4.9, were deployed in five of the six vehicles that successfully completed the course.

FIGURE 4.8 The Team Overbot with a 2D LIDAR.

FIGURE 4.9 A vehicle equipped with a 3D Velodyne LIDAR.

(a)

(b)

Working Principles

A LIDAR unit continuously emits pulsed beams of laser light, instead of the millimeter radio waves in RADAR, which bounce off objects and then return to the LIDAR sensor. A 2D demonstration of how LIDAR detects and measures the shape and topography of an object is shown in Figure 4.10. The laser source (blue square in the middle image) continuously rotates and emits a laser. The sensor detects a reflection of the laser from a surface and the LIDAR software and generates a *point cloud* in a representation of 3D space where the reflections are determined to have taken place. As shown in Figure 4.10 (bottom), this point cloud is consistent with the rectangular shape of the room, except where the LIDAR beams are reflected by the circular object. By continuously and smoothly connecting the points through specifically developed software or algorithms, the object shape can be determined, and the object can be recognized. Since objects have irregular shapes with varying reflective properties, the incident light beam gets scattered, and only a small fraction of the original beam returns to the LIDAR sensor. To account for the reduction in laser signal, a collimated laser source is used along with focused optics.

FIGURE 4.10 A 2D demonstration of a LIDAR measurement.

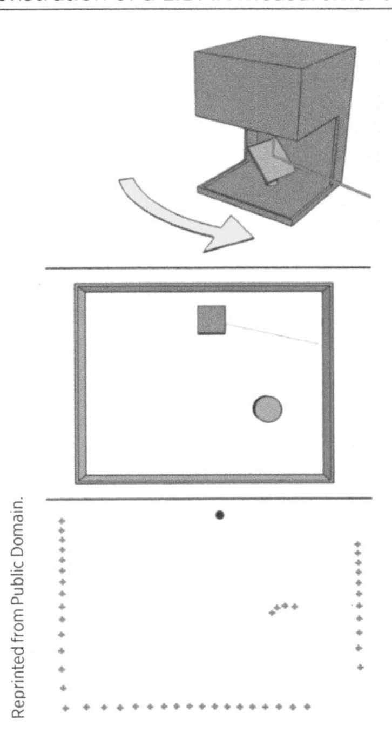

The distance to each reflected point is dictated by the equation $d = vt/2$, where $v = c$, the speed of light. The principle in CAV application is shown in Figure 4.11.

FIGURE 4.11 LIDAR measurement and application in a CAV.

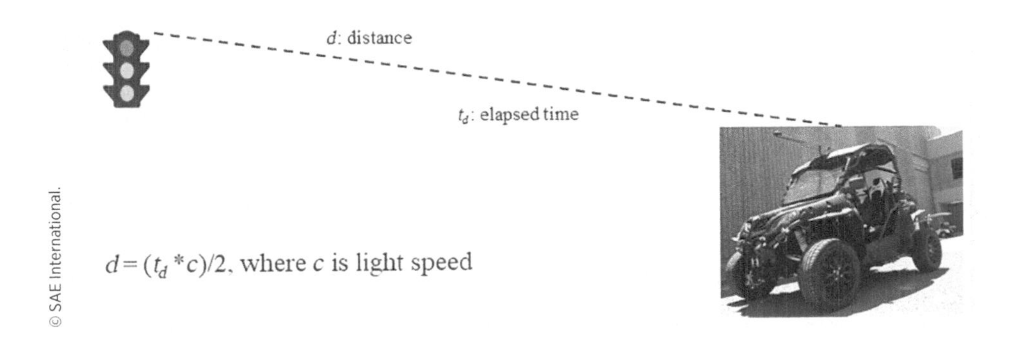

Types of LIDAR

LIDAR units can adopt different types and technologies summarized in Figure 4.12. Electromechanical units were initially the most common. These continuously rotate the emitter to generate 360-degree scans. Since they contain moving parts, electromechanical units are susceptible to perturbations due to road harshness and vibrations. A Micro-Electromechanical mirror operates using a similar principle, but in a smaller form factor. Flash and phased array LIDAR units are solid state and employ different strategies to emit

FIGURE 4.12 Different LIDAR units under development.

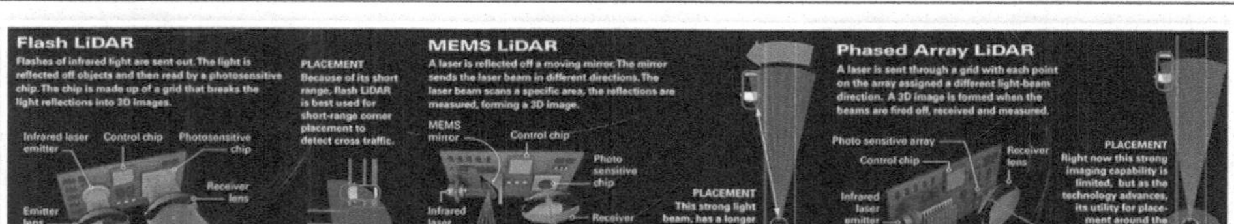

and collect signals in the different light-beam directions. According to Frost and Sullivan, over 90% of CAVs currently in the development stage use solid-state LIDAR units, which are smaller, cheaper, require less power, and more robust to noise and vibration than rotating, electromechanical units.

Characteristics

Electromechanical LIDAR units rotate to provide a 3D representation of the surrounding environment. Arrays of solid-state LIDAR units with different orientations can achieve the same functionality. Since the wavelength of a laser beam is much shorter than that of a RADAR, the resolution of the obtained 3D map based on the obtained point cloud will be generally much higher than if using RADAR measurements. The distance to any objects can be often be measured to within a few centimeters, and the detection range of most LIDAR units available today is about 200 m. As an active sensor, LIDAR can work well in different environmental conditions, including bright sun, low light, snow, and fog.

The main drawback of a LIDAR is the high cost compared with the camera and RADAR. Equipping a CAV with LIDAR can add costs in the tens of thousands of dollars. Electromechanical LIDAR experience high-speed rotations and may be bulky, fragile, and have low durability or reliability. Moreover, as the wavelength of laser beams used in LIDAR is very short and cannot penetrate materials, there is a vulnerability to occlusions. Since LIDAR is an active sensor, the potential issue of interference with other sources (e.g., other LIDAR units in the area) may cause concern. The construction of a 3D environment from LIDAR data imposes a significant computational burden and perhaps also a substantial energy cost.

In sum, LIDAR has the unique ability among the sensors discussed thus far to detect and construct a high-resolution 3D representation of the environment, which is needed for CAVs to make safe and reasonable decisions on the road. Many limitations of LIDAR units available today will likely be improved or partially resolved through new technology development. For example, solid-state LIDAR units are much cheaper, less fragile, and more compact than the large electromagnetic units driving early CAV development. These solid-state units can be made compact enough to be embedded into vehicle pillars and bumpers. However, the FOV of solid-state units is less than 360°, as shown in Figure 4.13. Thus many solid-state units may be required to cover the same FOV as that of a rotating, electromechanical LIDAR. Some of the advantages and disadvantages of LIDAR sensors are shown in the spider chart in Figure 4.14.

FIGURE 4.13 A flash LIDAR sensor with diffused light.

FIGURE 4.14 Spider chart of LIDAR sensors.

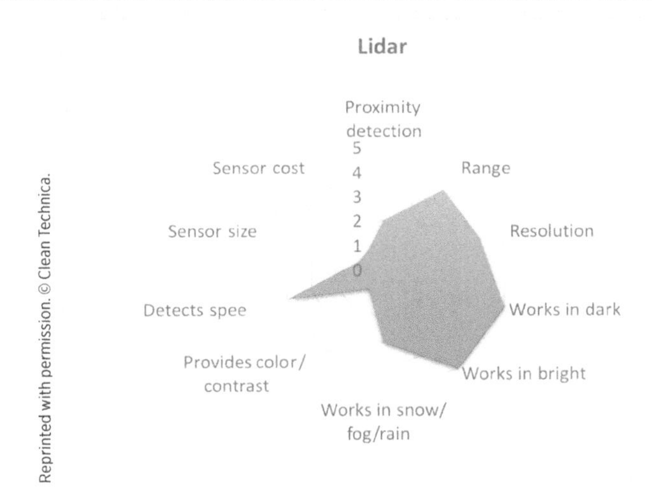

Ultrasonic SONAR

Definition and Description

SONAR stands for Sound Navigation And Ranging; ultrasonic SONAR is a subset of SONAR that uses sound frequencies in the ultrasonic range, i.e., not audible to humans. Ultrasonic SONAR emits and collects signals, much like RADAR and LIDAR. But as implied by its name, ultrasonic SONAR employs only sound waves for echolocation. This functionality was inspired by nature from the navigation of bats in the dark or dolphins in the sea, and the basic operating principle is shown in Figure 4.15. The first man-made SONAR devices were likely developed as a result of the Titanic sinking of 1912, and were first deployed in World War I.

FIGURE 4.15 Emitted and reflected sound waves of a SONAR unit.

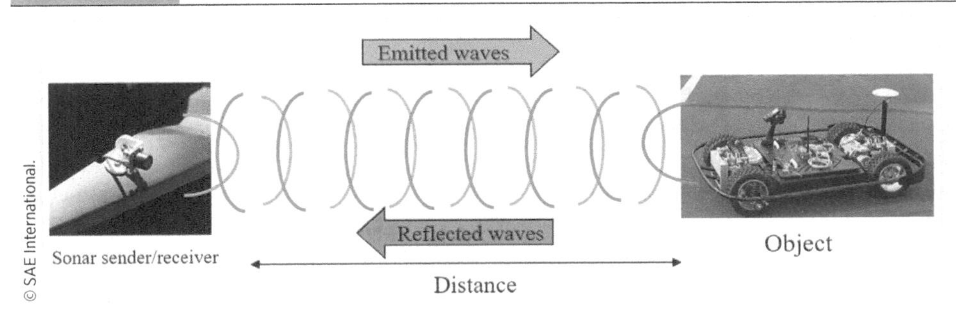

Sound waves travel better in water (e.g., in rain or snow) than do most electromagnetic waves. In the air, sound waves travel at 340 m/s, orders of magnitude lower than the speed of light. Lower wave propagation speeds mean that high-speed signal processing is not necessary, which can make SONAR less expensive to integrate into CAVs. However, the low wave propagation speed means that high-speed object tracking is not possible. SONAR sensors may be either active, with both transmitting and receiving modules, or passive, with only receiving modules.

Characteristics

Ultrasonic SONAR is most effective at a distance of approximately 5 m or less, and therefore is primarily used for near-vehicle monitoring. Ultrasonic SONAR has high directivity and is well suited for situations that require precision and tight maneuvers. Automatic parking is an example of a feature that can make use of these sensors, which are usually installed in the front and rear bumpers to provide overlapping FOVs, as shown in Figure 4.16, in order to identify the relative position of nearby objects.

FIGURE 4.16 Application of ultrasonic SONAR sensors in automatic parking.

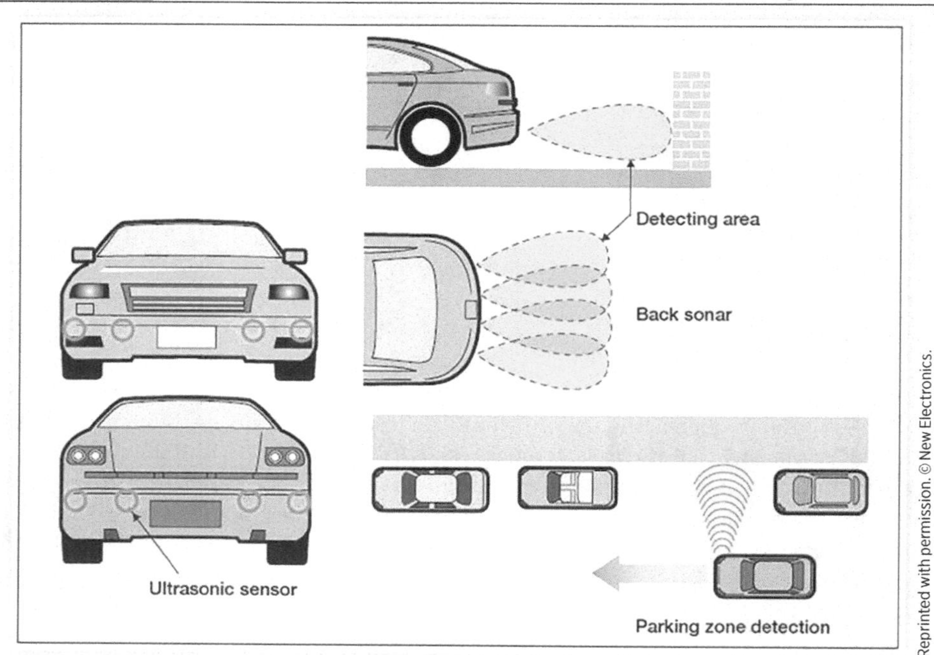

Among the sensors discussed in this chapter, ultrasonic sensors may offer the best performance in inclement weather such as rain or snow. Ultrasonic sensors are also rather inexpensive, costing as little as $20, and offer a measurement resolution on the order of 0.5 cm. Due to these advantages, ultrasonic sensors have already been widely employed on today's vehicles to support parking assist features, as discussed above. However, the short measurement range of ultrasonic sensors limits their application to low-speed maneuvers or nearby object detection. As with other active sensors, interference with other sources is a concern.

The spider chart of ultrasonic shown in Figure 4.17 summarizes the strengths and weaknesses of ultrasonic sensors. Despite poor range and resolution, their low cost, ability to detect nearby objects with high resolution, and low computational burden for processing mean that these sensors will likely continue to be an important component of CAV sensor suites.

FIGURE 4.17 Spider chart of ultrasonic SONAR sensors.

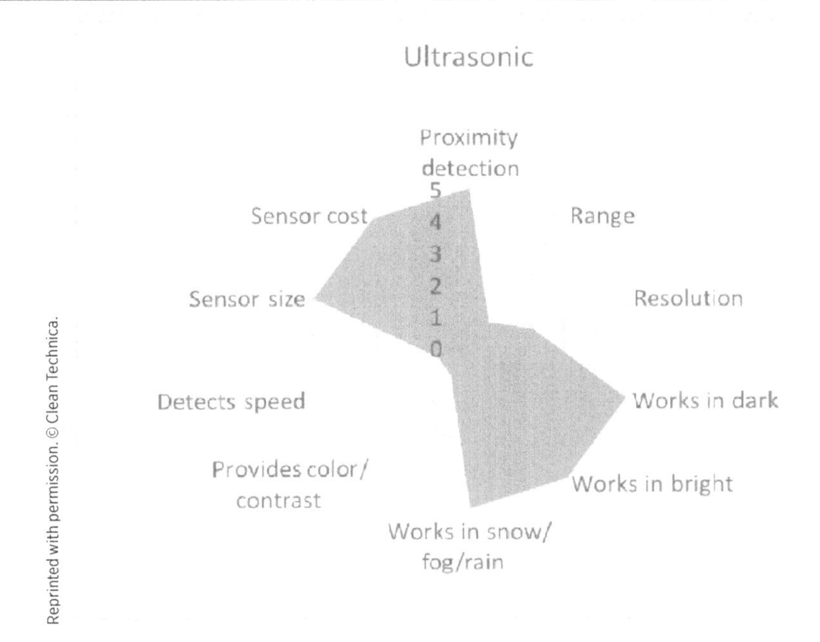

Other Important Sensors and Measurement Sources

While the aforementioned sensors represent the most common and useful sensors employed in CAVs, they may benefit information from the environment from other sources and sensors that can help ensure safe and efficient vehicle performance.

HD Maps

For the past several years, app-based navigation software like Google Maps or Apple Maps has been invaluable for personal travel and transportation. While their resolution and accuracy suffice for personal travel, and turn-by-navigation and lane guidance are enough to help guide human drivers to their destinations, CAVs may require higher-fidelity map data for localization and trajectory planning purposes.

As discussed in Chapter 2, HD maps encode information centimeter-level data that is far more detailed than what is available on app-based maps on smartphones, including information that most humans can easily perceive or take for granted. This may include data on the width of the lane and roadway, the locations of curbs and barriers, nearby traffic control devices, and the curvature of the roadway ahead. They may even encode information that may otherwise be ambiguous, such as where to stop if there is no stop bar at an intersection. HD maps are constructed using data from many of the sensors described earlier in this chapter, including LIDAR, RADAR, and cameras. Aerial imagery can also aid in the construction of HD maps.

High-Precision GPS

Today's vehicles and smartphones include GPS receivers to help with localization and navigation, but this hardware is generally insufficient for CAV applications.

Higher precision GPS units that can provide centimeter-level accuracy can improve CAV localization, making the detailed information encoded in HD maps more useful. A differential GPS, which communicates between the vehicle, a base station, and satellites in the sky, can achieve this level of accuracy. But differential GPS units are expensive, in the tens of thousands of dollars, making them cost prohibitive for most applications; notably, however Local Motors' automated shuttle Olli deployments currently use differential GPS as the routes are well defined and remain close to the base station. Further uninterrupted communication with a base station and satellite is often impractical in environments such as tunnels or dense, downtown urban areas. As a result, high-precision localization is usually achieved using sophisticated sensor fusion methods, which will be discussed Chapter 6.

Sensor Suites

Overview

A sensor suite is simply a set of complementary sensors equipped on a vehicle that provide the data needed for successful CAV operation. Based on the characteristics of each sensor described in this chapter, it may seem straightforward to integrate all the sensors, using the strengths of one to mitigate the weaknesses of another. Yet, at present, there is no single sensor suite to which CAV developers have converged. This is due to the continued improvements being made to each sensor modality, the costs associated with equipping large amounts of sensors, and new data processing methods and sensor fusion methods that can glean more useful data from the available sensors.

To present one example, a typical CAV with various sensing systems for different driving purposes is shown in Figure 4.18. Note that the locations, numbers, and models of these sensors, such as cameras, RADAR units, LIDAR units, and ultrasonic sensors can vary as well. Some CAV developers have dramatically changed their sensor suite over time. When Uber moved from the Ford Fusion to Volvo XC90 vehicle platform for its ADS, it reduced the LIDAR count from seven (a 360° spinning, roof-mounted LIDAR sensor plus six other LIDAR sensors) to one (just the roof-mounted LIDAR sensor). Even the use of LIDAR itself is not universally accepted. Though most CAV developers include one or more LIDAR units in their sensor suites, Tesla notably does not. At a 2017 TED event, Tesla CEO Elon Musk pointed out that humans carry out the driving task equipped with what are essentially two cameras, stating:

> There's no LIDAR or RADAR being used here [referring to Tesla's ADS]. This is just using passive optical which use essentially what a person uses. The whole road system is meant to be navigated with passive optical or camera, so once you solve cameras, or vision, then autonomy is solved. If you don't solve vision, it's not solved. That's why our focus is so heavily on having a vision neural net that's very effective for road conditions. You can absolutely be super-human with just cameras. You could probably do ten times better than humans with just cameras.

FIGURE 4.18 Sensors for different purposes.

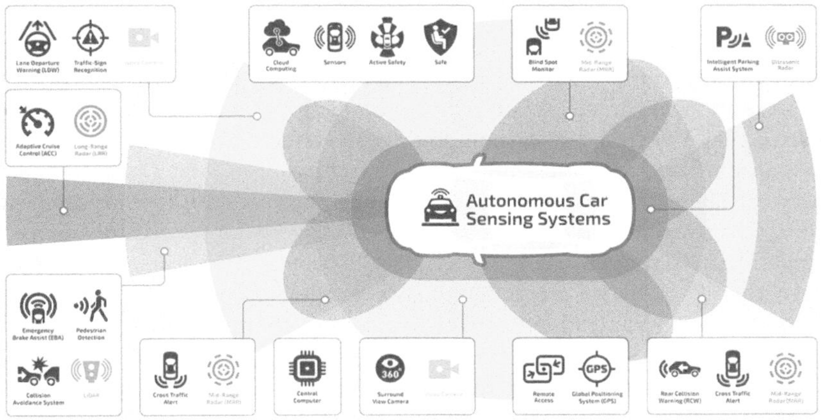

Autonomous Car Sensing Systems

iStock.com/Yuriy Bucharskiy.

Sensor Suite: Functionality

Since there is no standard on which sensors to include, how many sensors are needed, and where to mount the sensors on a vehicle, the suitability of a sensor suite for CAVs is determined by assessing its ability to support core CAV functionality. Specifically, one can establish, step by step, whether each of the safety functions of CAVs is suitably enabled by the selected sensors and data fusion methods, as shown in Figure 4.19, which also lists some common sensor requirements that are listed on the upper left corner. Note that for a practical application, the vehicle's physical dimension and shape will also impact sensor allocation and selection. Optimization of the sensor suite for cost and function can be achieved once verification of the required functionality is complete.

FIGURE 4.19 Sensor requirements and functions.

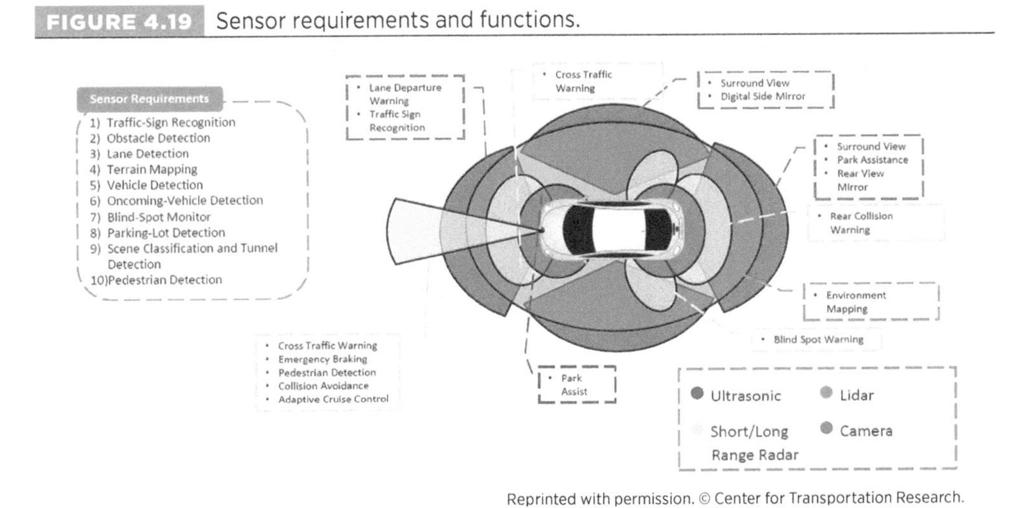

Reprinted with permission. © Center for Transportation Research.

Actuation and Propulsion Hardware

While sensors, sensing hardware, and data processing for CAVs have received significant attention and investment, vehicle actuators are less widely discussed. This is because the former presents the far greater set of technical challenges and barriers to CAV deployment. In the same vein, vehicle actuation and propulsion systems, including ICEs, EMs, steering systems, and brakes, are much more mature technologies than the sensing hardware discussed earlier in this chapter. It is likely that existing actuation systems can be incorporated into CAVs with relatively little modification. By replacing the human steering and pedal inputs with control signals from an ADS, CAV control could be achieved via the same types of actuation hardware present on today's vehicles.

Nevertheless, CAVs are likely to take advantage of actuator technologies that have been recently developed not for automated driving, but to improve the responsiveness of existing vehicle controls, or to facilitate the control of electrified vehicles. A few of these enabling actuator technologies are described below. More discussion on actuation hardware and techniques is contained in Chapter 7.

Steer-By-Wire

ICEs had traditionally been controlled by a throttle pedal, which modulated a throttle valve that regulated the supply of air to the engine. Electronic throttle control, sometimes called throttle-by-wire, replaced this mechanical link with an electronically controlled device that regulated the throttle position based on both the pedal input and a number of other measured variables. Electronic throttle control is necessary for features such as cruise control, traction control, and stability control.

Likewise, steer-by-wire offers the same functionality by removing the mechanical coupling (steering column) between the hand steering wheel and vehicle's wheels. Developed to improve the responsiveness and on-center steering control for human-driven vehicles many years ago, steer-by-wire uses a controller to determine the appropriate wheel steer based on the position of the driver's hand wheel and several measured variables, as shown in Figure 4.20. A straightforward application of steer-by-wire to CAVs would simply replace the driver's hand wheel angle with an equivalent input from the ADS.

FIGURE 4.20 Schematic of a steering by wire system.

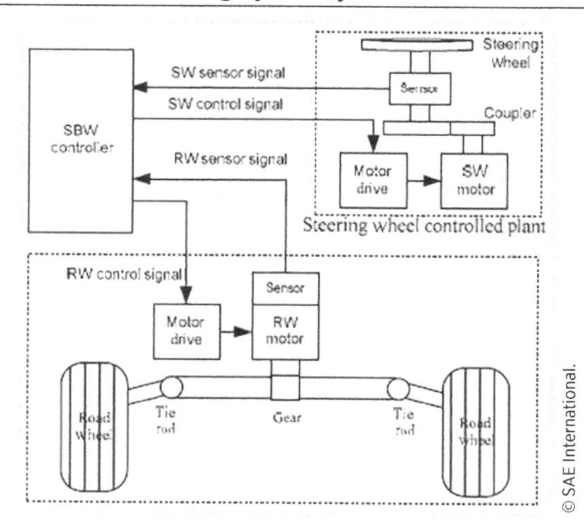

However, lingering safety and reliability concerns surrounding steer-by-wire systems have prevented their widespread market adoption following its introduction on the Infiniti Q50 in 2013. Thus, most CAV developers currently use platforms with traditional steering columns that incorporate an EM to steer the wheel automatically, as shown in Figure 4.21. As CAVs are still being evaluated and tested, often with human safety drivers on board as a backup measure, the use of traditional steering hardware means that these safety drivers can take control when needed, but otherwise let the ADS fully control the vehicle.

FIGURE 4.21 Automatic steering through an EM.

A Controller Area Network (CAN) bus is an enabling and fundamental technology to communicate short messages among different vehicle embedded systems, including but not limited to steer-by-wire systems, powertrain systems, and safety systems. A CAN bus, compared with other communication technology, such as analog signals, can achieve more robust, faster, and safer communication through simple hardware or wiring setups. Typically, a backbone configuration with two CAN bus wires, CAN high and CAN low, can be utilized to communicate sensing, control, and diagnostic information at 250k or 500k bit/sec rates.

Rear-Wheel Steering

Rear-wheel steering was studied and developed three decades ago, but never found favor on commercially available passenger vehicles. Rear-wheel steer, shown in Figure 4.22, can be controlled in the same manner as front-wheel steer, with steer-by-wire potentially reducing the necessary mechanical complexity. While adding rear-wheel steer can introduce unstable dynamics, especially at higher speeds, systems that offer both front- and rear-wheel steer (known as four-wheel steer, or 4WS) can achieve improved lateral stability and low-speed maneuverability. The latter may be especially advantageous for low-speed last-mile CAV shuttles that operated in dense, urban environments. While the cost of 4WS is a barrier

FIGURE 4.22 A rear-wheel steering system in a CAV.

to adoption on conventional, human-driven vehicles, the incremental cost is much lower on more expensive CAVs, and may be worth the expense in certain CAV applications.

Electric Propulsion and In-Wheel Motors

Electric propulsion systems can take many different forms. While CAVs need not have electric powertrains, the use of electrified platforms for CAVs offers greater opportunities for CAVs to increase the energy efficiency of our entire transportation system. In addition to offering much higher thermal efficiency than ICEs, EMs are also well suited to precise speed and torque control, allowing the CAV to react more quickly and effectively to safety threats.

One type of electrified architecture that offers unique benefits for CAVs is the in-wheel motor. In contrast to centralized propulsion motors that distribute power to the wheels through axles and shafts, in-wheel motors offer local, distributed torque generation at each wheel. The Olli 2.0 automated, electric shuttle by Local Motors uses in-wheel motors (the 1.0 used a central motor with transmission). In-wheel motors, also called hub motors, are named or called based on the location of EMs. In principle, different types of EMs can be put on the wheels to provide driving and braking torques. Practically, considering the power requirements and compact spaces in the wheels or hubs, two types of EMs are often used as in-wheel motors, namely, brushless direct current (BLDC) and induction motors. BLDC is a type of AC motor with power or torque properties similar to DC motors. Compared with the common DC motors, the feature of no brushes (commutators) makes the BLDC maintenance easier. Induction (or asynchronous) motor is one type of AC EM, in which the current to generate torques in the rotor is obtained by electromagnetic induction from the magnetic field of the stator winding. Both BLDC and induction motors have the characteristics of high torques at low rotational speeds, which make them well suitable for vehicle driving applications.

A vehicle with four in-wheel motors offers new opportunities to distribute power and torque generation to achieve precise control and efficient propulsion. Together with 4WS actuation, as described before, a CAV with four in-wheel motors will have redundant actuators together with various complementary sensors. Thus, benefiting from both redundant sensors and actuators, CAVs can have safer, more energy-efficient, and more agile motions with the new actuation technology, as initially demonstrated in Figure 4.23.

FIGURE 4.23 A CAV driven by four independent in-wheel motors.

References

Barnard, M. (2016, July 29). Tesla & Google Disagree About LIDAR—Which Is Right? Retrieved from CleanTechnica.com: https://cleantechnica.com/2016/07/29/tesla-google-disagree-lidar-right/

Davies, C. (2017, August 18). Self-Driving Car Tech Just Took a Big Step Closer to Mainstream. Retrieved from Slashgear.com: https://www.slashgear.com/self-driving-car-tech-just-took-a-big-step-closer-to-mainstream-18495775/

Du, Z., Wu, J., He, R., Wang, G., Li, S., Zhang, J., & Chen, G. (2021). A Real-Time Curb Detection Method for Vehicle by Using a 3D-LiDAR Sensor. SAE Technical Paper 2021-01-0075. doi:https://doi.org/10.4271/20201-01-0076

Kuhr, J. (2017). Connected and Autonomous Vehicles: The Enabling Technologies. *The 2017 D-STOP Symposium*. Austin, TX. Retrieved from https://www.slideshare.net/ctrutaustin/connected-and-autonomous-vehicles-the-enabling-technologies

Lim, K., Drage, T., Zheng, C., Brogle, C., Lai, W., Kelliher, ... Braunl, T. (2019). Evolution of a Reliable and Extensible High-Level Control System for an Autonomous Car. *IEEE Transactions on Intelligent Vehicles*, 4(3), 396-405.

Mandlburger, G., Pfennigbauer, M., Schwarz, R., Flory, S., & Nussbaumer, L. (2020). Concept and Performance Evaluation of a Novel UAV-Borne Topo-Bathymetric LiDAR Sensor. *Remote Sensing*, 12(6), 986. doi:10.3390/rs12060986

New Electronics. (2010, May 12). An Introduction to Ultrasonic Sensors for Vehicle Parking. Retrieved from https://www.newelectronics.co.uk/electronics-technology/an-introduction-to-ultrasonic-sensors-for-vehicle-parking/24966/

Team Overbot. (2005). DARPA Grand Challenge 2005 Technical Paper.

Wang, D., Watkins, C., & Xie, H. (2020). MEMS Mirrors for LiDAR: A Review. *Micromachines*, 11(5), 456. doi:10.3390/mi11050456

Wikipedia.org. (n.d.). Lidar. Retrieved August 31, 2021, from Lidar: https://en.wikipedia.org/wiki/Lidar

Yao, Y. (2006). Vehicle Steer-by-Wire System Control. SAE Technical Paper 2006-01-1175. doi:https://doi.org/10.4271/2006-01-1175

VOL
MAP
RADIO
MEDIA
NAV
FRONT
REAR
PASSENGER
AIR BAG
A/C
CLIMATE
MODE
DUAL
CleanAir
TEMP
AUTO
TEMP

5

Computer Vision

Humans can intuitively understand a traffic scene while driving a vehicle on the road by looking at the surrounding environment. Yet how the human visual system works is still not well understood. Interestingly, it is not the structure and function of the eye but the brain and neurons (especially the visual cortex) that puzzle the researchers (Szeliski, 2010). For a CAV, the perception system is its "visual cortex." It contains various computer vision algorithms to extract useful information from the sensor input and construct an intermediate representation of the surrounding environment. These intermediate representations eventually allow the vehicle to make essential decisions to drive itself on the road safely.

With the advancements in machine learning (ML) and especially deep learning, many visual perception problems can be solved in a modern way through data-driven models. Also, thanks to the exponential increase of computational power of a single computer chip, especially the Graphics Processing Unit (GPU), these solutions can be deployed on the vehicle and run in real time. In this chapter, an overview of computer vision for CAV is provided. Due to the nature of visual perception sensors, the subject in computer vision study can be either 2D structures, i.e., images and videos, or 3D structures, i.e., depth images, voxels, or 3D point clouds. Hence, in section "Image and 3D Point Cloud," the basics of these structures and algorithms to process them at a low level from a computer science and signal processing perspective are shown. Next, in section "Deep Learning," the basics of deep learning, which is the key to make the perception system of a CAV possible and achieve a performance comparable to humans, are illustrated. Equipped with these fundamentals, in section "Perception Tasks for CAV," eight critical visual perception tasks in traffic scene understanding for automated driving are explored. Finally, in section "Perception System Development for CAV," practical issues in developing the visual perception system of a CAV are discussed, and a case study is presented. As with all chapters in this book, computer vision itself is a big topic worth an entire book of its own. This chapter mainly serves as an introduction of this subject in the context of the CAV without much detail on the know-how or the implementations.

Image and 3D Point Cloud

The inputs of a perception system are typically the outputs of sensors, and in the CAV case, the camera and LIDAR. Hence, mathematical representations of these inputs and outputs need to be formulated, and in this section, the representations, i.e., images and 3D point clouds, will be discussed.

Image Formation

An image is usually formed by capturing the light emitted or reflected from objects and the environment in the 3D world through an optical lens and an image sensor (i.e., a camera), as shown in Figure 5.1. The image sensor is typically arranged in a grid to evenly sample the image plane of the optical lens, and each cell of the grid is called a pixel. Mathematically, an image is represented as a 2D matrix, and each element of the matrix contains some information of the captured light at that spot (i.e., the pixel value). If each pixel has only one value for the light intensity, it is a grayscale image. This type of image is common for cameras working at a range of light spectrum not visible to human eyes, e.g., an infrared camera. To obtain the color information in a way similar to human perception, each pixel of the camera is covered by three types of color filters on the image sensor (i.e., red, green, and blue) to record the light intensity of each color channel individually. Hence, there are three color values (R, G, B) associated with each pixel, and typically each value is an 8-bit number from decimal 0 to 255; (0, 0, 0) is black, (255, 0, 0) is red, and (255, 255, 255) is white. Such an image is an RGB image, or just commonly called a color image. A grayscale image can also be considered as an image of only one color channel. If a camera can capture images at a regular time interval, as most cameras can, this sequence of images is a video. Hence, from a mathematical perspective, a video is a 3D tensor made of individually sampled 2D matrices of images along the timeline. Images or videos are usually stored in a compressed format to save disk space. When they are processed or viewed, a program loads and uncompresses them into the computer memory in the format of multiple dimensional arrays like matrices or tensors. The sizes of such arrays are determined by the number of pixels in each row and column, known as the image resolution, e.g., 1920 * 1080.

FIGURE 5.1 Image formation.

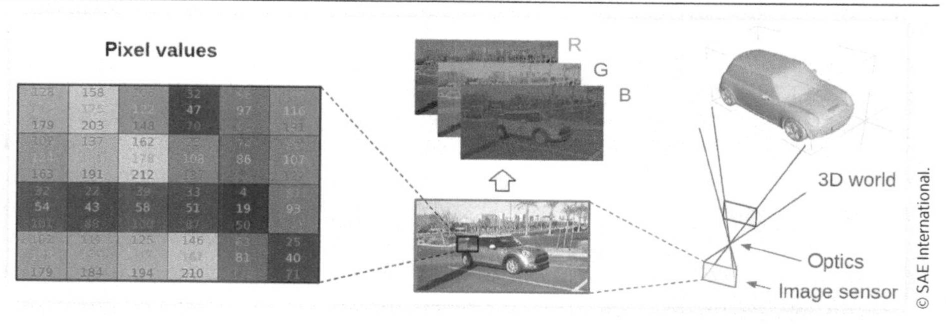

To explain the math and physics of image formation, a theoretical perspective is taken and the image sensor is considered to be a mathematical 2D plane in the 3D space, i.e., the image plane. Geometrically, each point on this plane is related to a collection of 3D points (typically in a line) by the camera projection, as illustrated in Figure 5.1 (right). However, in reality, the image sensor and the imaged objects in the 3D world are on different sides of the lens; here the image plane and the imaged objects are both placed on the frontal side of the lens for convenience (they are mathematically equivalent). Also for convenience, the camera is assumed to have its own reference frame in 3D where the origin is the optical center of the lens (i.e., the point that every ray of light converges to the same point in Figure 5.1), the x-axis is along with the rows of the image, the y-axis is along with the columns of the image, and the z-axis points to the frontal side of the camera. Most cameras can be modeled as a pinhole camera, where the camera projection is a set of linear equations (represented as matrix-vector multiplication), as shown in Figure 5.2. Assuming a 3D point X has coordinates (X, Y, Z) in the world reference frame, it can be written in homogeneous coordinates $(X, Y, Z, 1)^T$. Similarly, the corresponding 2D point x in pixel coordinate (u, v) on the image plane can also be written as $(u, v, 1)^T$ in homogeneous coordinates. Here both u and v are real numbers instead of integers, i.e., they are "subpixel" coordinates with infinite precision (in theory). All 2D and 3D points can be represented in this way as column vectors, and the pinhole camera projection of a point is $x = PX$, where P is a 3×4 matrix.

FIGURE 5.2 An illustration of the pinhole camera projection.

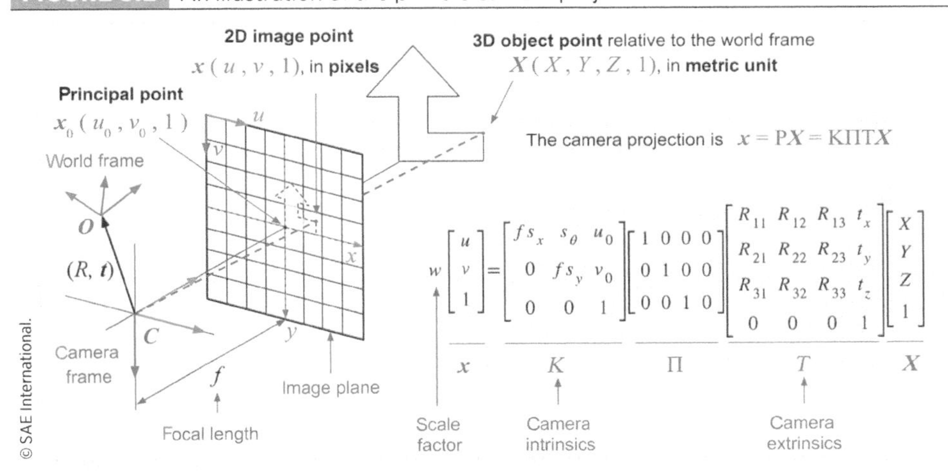

$$w\begin{bmatrix}u\\v\\1\end{bmatrix}=\begin{bmatrix}fs_x & s_\theta & u_0\\0 & fs_y & v_0\\0 & 0 & 1\end{bmatrix}\begin{bmatrix}1&0&0&0\\0&1&0&0\\0&0&1&0\end{bmatrix}\begin{bmatrix}R_{11}&R_{12}&R_{13}&t_x\\R_{21}&R_{22}&R_{23}&t_y\\R_{31}&R_{32}&R_{33}&t_z\\0&0&0&1\end{bmatrix}\begin{bmatrix}X\\Y\\Z\\1\end{bmatrix}$$

Specifically, this camera projection matrix contains three independent pieces of information, i.e., P = KΠT. They are the three steps of converting the 3D points to the camera reference frame from the world reference frame, projecting from 3D to 2D, and converting 2D points in the image plane to 2D pixels. The first piece of information is the camera extrinsics, or the camera extrinsic parameters, represented by the matrix T. The origin of the world reference frame is assumed to be at point O, the center of the camera is point C, as illustrated in Figure 5.2. The 3D Euclidean transformation from the world reference frame to the camera reference frame is represented by (R, t), a 3×3 rotation matrix R and a 3D translation vector t. The camera extrinsics are essentially this transformation, which also provides the relative pose of the camera in the world. Since it can change based on the camera position and orientation, it is extrinsic to the camera. For a camera mounted on a

CAV, its extrinsics are obtained from the vehicle location relative to an HD map for automated driving (see Chapter 2). The second piece of information is the normalized pinhole projection, represented by the matrix Π, which is essentially a projection from the 3D world using a normalized pinhole camera with a focal length of 1 metric unit and pixel size of 1×1 metric unit. If the lens has distortion, Π will be a nonlinear mapping instead of a matrix. In practice, a distorted pinhole projection can be approximated using a polynomial function with a few extra parameters for a specific lens. The third piece of information is the camera intrinsics, or the camera intrinsic parameters, represented by the matrix K. This matrix maps normalized image plane coordinates to pixel coordinates specific to the image sensor and lens, which is intrinsic to the camera itself and irrelevant of its pose. Typically, the camera intrinsics contain five parameters $(u_0, v_0, fs_x, fs_y, s_\theta)$. Here (u_0, v_0) means the pixel coordinates of the principal point, i.e., the intersection of the principal axis of the lens (the z-axis of the camera frame), f is the focal length of the camera, (s_x, s_y) are the number of pixels per metric unit in the x-axis and y-axis of the camera, and s_θ is the pixel skewness factor. For a typical camera whose pixels are squares in a grid, s_θ is always zero, and s_x is equal to s_y. Otherwise, s_x is different from s_y if the pixels are rectangular instead of square (which is rare), and s_θ is nonzero if the pixels are in parallelograms (the x-axis and y-axis of the image plane are always perpendicular, but the pixels are not rectangular and the columns are not aligned with the y-axis, which is also rare). For a camera mounted on a CAV, its intrinsics are calibrated in the lab or the factory using a set of known 3D-2D point correspondences.

Given these three pieces of information, any geometric structures in the 3D world can be projected onto an image with a camera, and any point or pixel on an image can be back-projected to a ray in 3D. During the projection in either direction, a scale factor along the z-axis of the camera is lost (i.e., the depth, shown as "w" in Figure 5.2). Hence, for a regular RGB camera, the depth of a specific pixel or the actual 3D scale of the line connecting two pixels cannot be determined given a single image. However, there is another type of camera, generally called RGB-D camera ("*D*" means depth), that can provide the depth of each pixel. The resulting image is an RGB-D image with an additional "color" channel for the depth values, or sometimes that channel is used independently as a grayscale depth image because it may have a different resolution from the RGB image. Such a camera usually actively projects beams of invisible light (typically infrared) with encoded spatial pattern (e.g., structured-light camera) or temporal pattern (e.g., time-of-flight (TOF) camera) and captures the change of the spatial or temporal pattern of the reflected light to decode the depth. They are similar to a LIDAR, but they generally do not have mechanically moving parts for scanning (sometimes they are called "scanless LIDAR"). Yet these sensors do not work well in bright sunlight, and they are rarely used for automated vehicles driving on the road. Still, the concept of the depth image is useful since it may be derived by stereo triangulation (see section "3D Depth Estimation") or projecting a LIDAR scanning onto an image taken by an RGB camera at the same time (in this case, it is a sparse depth image with valid depth values for only a limited number of pixels since the angular resolution of a LIDAR cannot match a camera).

Geometric image formation is only half the story; the second half is photometric image formation (i.e., lighting, shading, and optics), which relates points and lines to light intensity and color values. As a CAV does not need to recover the precise interaction of light and surfaces of surrounding objects as well as the environment for graphical simulation, the photometric image formation process will only be discussed briefly. The brightness of any pixel is determined by the amount of light projected onto that pixel, which can be traced back to a light source directly without reflection or indirectly through one or multiple surface reflections and scattering. The reflection is generally modeled as a mixture of diffuse reflection and specular reflection. In the former case, the intensity of the reflected light is isotropic, and it depends on the surface reflectance or albedo, the angle of the incident light, and the color of the surface. In the latter case, the intensity of the reflected light is dominated by the angle of the incident light. Light scattering is generally modeled as an ambient illumination light source filled uniformly in the space without direction and reflected diffusely by every surface. Given the parameters of all light sources, surfaces, cameras, and additional parameters of the optics considering focus, exposure, aberration, etc., the color value of each pixel can be computed through a physical model, which approximates the photometric image formation procedure in reality.

Image Processing

The first stage of a visual perception system is usually a set of low-level processing to obtain better images or low-level information suitable for further processing or ML steps discussed in section "Perception Tasks for CAV." The image is a function $I(x, y)$ or just I for short, where x and y are the pixel coordinates equivalent to (u, v) and the function value is the value of the pixel at (x, y), as illustrated in Figure 5.3. In this figure, the y-axis is intentionally reversed and the origin is set to the bottom-left corner of the image. Similarly, a video can be considered as a function $I(x, y, t)$ with an additional input variable t representing the time for each video frame. The output of many image processing algorithms is another image, and it is modeled as a function $f(I) = I'$, which maps the pixel values of one image to new values. Some algorithms generate general-purpose, low-level structural descriptions such as corners, key points, lines, or other geometric shapes so that a more advanced problem-specific algorithm can use them. Many image processing algorithms also work on every single pixel $I(x, y)$ individually or apply a mathematical operation on a small local neighborhood $\{I(x, y) \mid x_1 < x < x_2, y_1 < y < y_2\}$, and hence, they can be implemented efficiently and run in parallel. Most algorithms are developed using signal processing approaches that are different from computer vision and ML approaches. A few commonly used image processing steps for visual perception in a CAV will be discussed.

FIGURE 5.3 An example of an image as a function.

The simplest image processing algorithms are those running the same operation at each pixel, such as brightness adjustment, contrast adjustment, color correction or white balance, changing the color range (e.g., from 8-bit integers to 32-bit floating-point numbers), converting color space (e.g., from RGB to Hue-Saturation-Value [HSV]), normalization (e.g., converting pixel values from [0, 255] to [−1, 1] using statistical results of the whole image), etc.

Three examples are shown in Figure 5.4. First, in Figure 5.4 (left), a nonlinear color mapping is applied to each pixel of each color channel of the image, where the input-output pixel value relation is denoted as a function $f()$ plotted as the curve on top of the image. It can be seen that the bright portion of the image is disproportionally brightened further so that the overall contrast of the image is higher. This kind of nonlinear mapping is commonly used for gamma correction when the image sensor or the display maps light luminosity nonlinearly to pixel values. Second, in Figure 5.4 (middle), the pixel values of the RGB image are converted to the HSV color space, where the hue is a number from 0 to 360, indicating a specific color on the color wheel. The image is then thresholded with a predefined region denoted as the three ranges above the image (i.e., if the value of a pixel is in that range, it is set to white, and otherwise, it is set to black). The body of the vehicle is selected where the color falls into the specified region (i.e., blue with high saturation and intensity). This kind of mask is useful for structural analysis of image patches or detection of certain color patterns, and this step runs much faster than a powerful neural network object pattern detector trained on tons of data. Third, in Figure 5.4 (right), histogram equalization is applied on the grayscale image to normalize the brightness. Specifically, given the histogram $h(j)$ of the pixel value of the grayscale image [where j means the jth histogram bin, $(0 \leq j < 256)$, a cumulative distribution

$$h'(i) = \sum_{j=0}^{i} h(j)]$$

can be calculated, the value of $h'(i)$ is normalized to [0, 255], and then it is used to remap the pixel values of the grayscale image to obtain the equalized image $I'(x, y) = h'(I(x, y))$. The histograms of pixel values before the normalization and after the normalization are shown in the plot above the image. This algorithm is usually used to reduce the influence of different lighting conditions of the camera and improve the robustness of later processing steps.

FIGURE 5.4 Examples of pixel-level image processing.

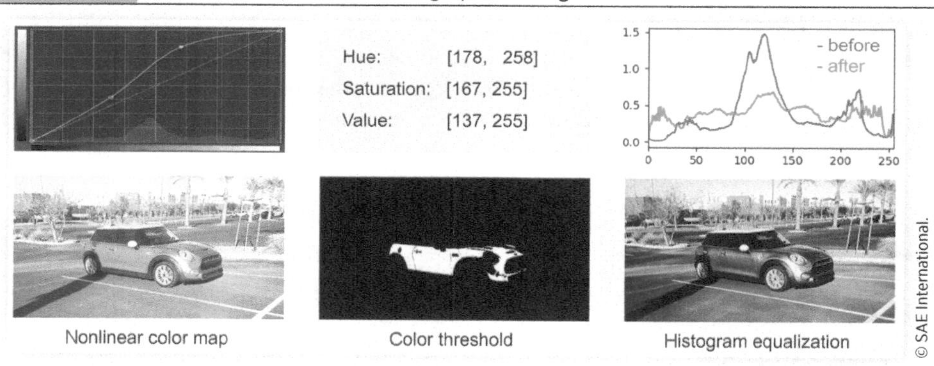

Besides individual pixel-level operations, a geometric transformation can be applied on an image, i.e., for any pixel at point $x\,(x, y, 1)$ in homogeneous coordinates on the original image $I(x, y)$, the transformed point $x'\,(x', y', 1)$ is at the place $x' = Hx$, and the transformed image $I'(x', y') = I(x, y)$. Here H is a 3×3 nonsingular matrix that characterizes the transformation. Depending on the degree of freedom in H, different types of transformation can be done for different purposes, as shown in Figure 5.5. One usage of this kind of geometric transformation on a CAV is the 360° parking assistance, also called around-view parking assistance, as shown in Figure 5.6. Typically, the vehicle is equipped with four cameras in four directions. Given the camera parameters obtained from factory calibration, a homography transformation between the image and the flat ground can be derived for each camera, and the images taken by these four cameras can be remapped and stitched together to show the surrounding of the vehicle from a bird's-eye-view on the display inside the vehicle to make the parking maneuver easier.

FIGURE 5.5 Examples of geometric image transformation.

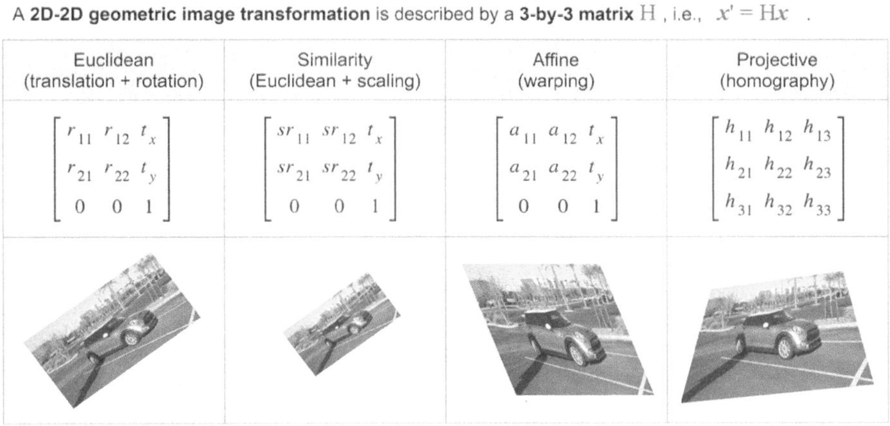

A 2D-2D geometric image transformation is described by a 3-by-3 matrix H, i.e., x' = Hx .
Euclidean (translation + rotation)
Similarity (Euclidean + scaling)
Affine (warping)
Projective (homography)
$\begin{bmatrix} r_{11} & r_{12} & t_x \\ r_{21} & r_{22} & t_y \\ 0 & 0 & 1 \end{bmatrix}$
$\begin{bmatrix} sr_{11} & sr_{12} & t_x \\ sr_{21} & sr_{22} & t_y \\ 0 & 0 & 1 \end{bmatrix}$
$\begin{bmatrix} a_{11} & a_{12} & t_x \\ a_{21} & a_{22} & t_y \\ 0 & 0 & 1 \end{bmatrix}$
$\begin{bmatrix} h_{11} & h_{12} & h_{13} \\ h_{21} & h_{22} & h_{23} \\ h_{31} & h_{32} & h_{33} \end{bmatrix}$

FIGURE 5.6 An example of around-view parking assistance.

Actual vehicle position
Around-view
An illustration of the camera coverage

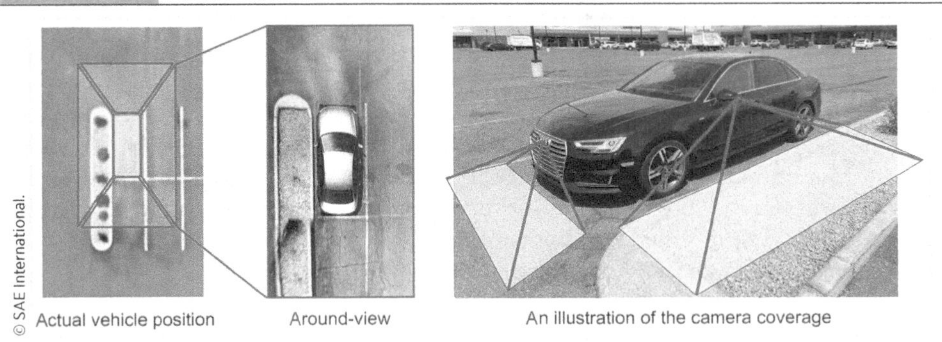

Another commonly used image processing algorithm is 2D convolution, which runs a neighborhood operator in the vicinity of a local group of pixels, as shown in Figure 5.7. In this processing procedure, a predefined convolutional kernel $K(i, j)$, typically an $m \times m$ matrix, where m is an odd number, e.g., 3×3, 5×5, etc., is applied on each $m \times m$ sliding window of the image with an element-wise Fused Multiply-Accumulate (FMA) operation. Depending on the size and the specific values of the kernel, a variety of desired results can be achieved. For example, in Figure 5.7, two examples are shown, i.e., the Gaussian blurring and the Laplacian edge extraction, applied on the image using the same convolution algorithm, but with different kernels. The values of the kernels are shown at the left-bottom corner of the resulting images. It is interesting to see that a slight change of the kernel can have drastic differences in the results, e.g., one for blurring the sharp edges and the other for extracting the edges. In fact, convolution is a fundamental operation for the much powerful and capable CNNs discussed in section "Deep Learning," where the difference is that hand-engineered kernels are used here while the neural networks learn appropriate kernels from data and cascade them with nonlinear activations to extract a plethora of image features at different semantic levels (and usually not understandable to humans).

FIGURE 5.7 An example of 2D convolution and image filtering.

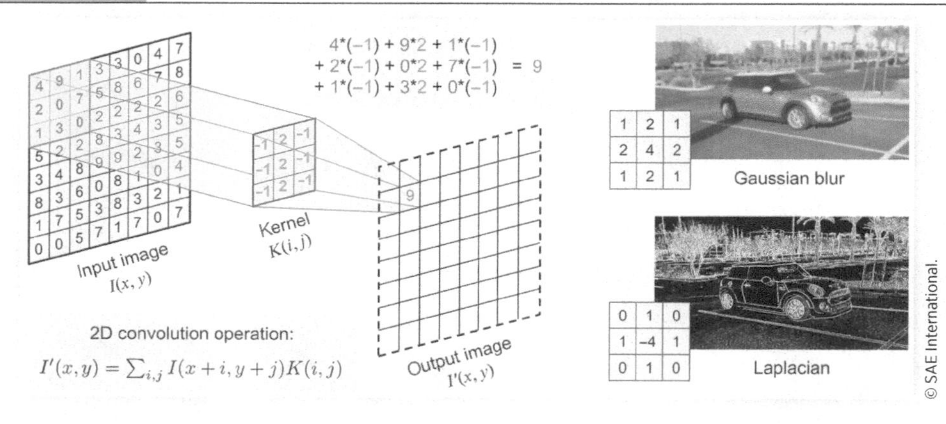

In many cases, the interest is in obtaining a whole processed image, but also a set of structural components on the image such as key points or feature points, edges, lines, contours, blobs or segmentations, etc. For example, in Figure 5.8, a few point correspondences on two different views of the same vehicle by a moving camera are shown. They are derived from detecting a type of feature point called the Oriented FAST and Rotated BRIEF (ORB) feature points and matching with the ORB descriptor of these features. If these features are landmarks, given enough number of good point correspondences from images at multiple different views, the relative poses of the camera at each view (i.e., visual odometry) can be recovered as a later processing step, which is very useful for a CAV to localize itself. With the relative poses of the camera, an even better idea is to triangulate the 3D position of these points and reconstruct the 3D shape if they are dense and on a rigid object (i.e., structure-from-motion). This is a way to obtain an HD map of a city by just manually driving a vehicle with cameras and other sensors, and as discussed in Chapter 2, such a map is crucial to allow a CAV to drive itself.

FIGURE 5.8 Examples of feature point detection and structural analysis.

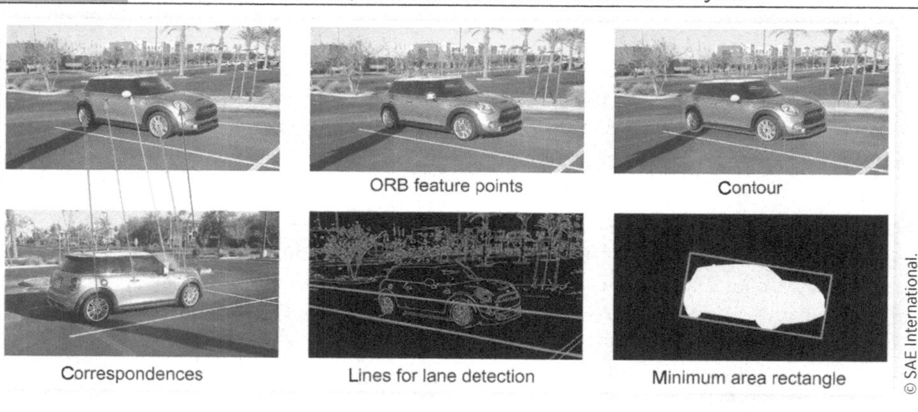

Even with a single image, a lot of interesting information about the structures in the 3D world can be obtained. For example, as shown in Figure 5.8 (middle), the Canny edge detector can be applied to the image and extract the lines to detect the lane boundaries.

Since the two sides of a lane are typically in parallel, they meet at a vanishing point. All vanishing points on the ground will reside on the horizon line, and the ability to detect them allows a CAV to understand its own 3D orientation relative to the ground. Moreover, if the vehicle can detect other vehicles and know the exact pixels on those vehicles on the image, i.e., the segmentation mask as shown in Figure 5.8 (right), it can calculate the contour of the mask by checking the pixel connectivity and fit a minimum area rectangle on the contour using a "rotating calipers" algorithm. Since most vehicles are "boxy," the lower edge of this rectangle tells us the orientation of that vehicle and the contacting points of the wheel and the ground. The 3D location and dimension of that vehicle can be further derived using the camera parameters, its pose relative to the ground, and the vanishing points (see the case study in section "Perception System Development for CAV"). These are all valuable pieces information for a CAV to make wise driving decisions in time. However, detecting other vehicles and calculating the segmentation mask promptly for each of them is a nontrivial task that requires deep learning and carefully engineered algorithms to achieve good performance (see sections "Deep Learning" and "Segmentation").

3D Point Cloud Formation

A 3D point cloud is a collection of 3D points. In this chapter, the 3D point cloud is mainly discussed in the context of a single real-time LIDAR scanning, which provides relatively structured data points in multiple laser beams from the same center with well-defined angular separation. For such a point cloud, each point consists of a timestamp, a 3D coordinate in the format of (range, azimuth angle, altitude angle) relative to the LIDAR itself, and a laser reflection intensity value. Similar to a camera, the LIDAR scan can only reach points in the line-of-sight and 3D information on an occluded surface cannot be obtained. For example, the Velodyne Ultra Puck LIDAR (Velodyne, 2019) has a 360° horizontal FOV with a roughly 0.1° resolution (~3,600 points in a circle) and 32 laser beams in the vertical direction (−25° to +15°, nonlinearly distributed). The points from a single scan from such a LIDAR can be put into a 3,600×32 matrix like an image in a range view as shown in Figure 5.9 (left), where each row indicates a single laser beam, each column indicates a specific azimuth angle, and the value of each element is the range as well as other attributes of that measurement. If a CAV has both a LIDAR and a camera, with their relative position and orientation, this laser range image can be projected to the camera image as an additional sparse depth image channel, so that a collection of computer vision algorithms on RGB-D can be applied. This is also a way of early fusion of LIDAR and camera data. Sensor fusion is discussed in more depth in Chapter 6.

Another commonly used representation is the bird's-eye view as shown in Figure 5.9 (middle), where the points are arranged according to the positions on the XOY-plane of the LIDAR reference frame. In this way, a set of vectors $S = \{(x_i, y_i, z_i, a_i)\}$ can be used to represent such a point cloud by ignoring the order of points and the timestamp. Here all the coordinates are in the reference frame defined by the LIDAR and "a_i" is a point attribute, typically the reflectance intensity value. Sometimes the points are quantized into individual "pillars," where a pillar contains all points whose (x, y) coordinates are within a grid cell on the XOY-plane. This quantized view can be further quantized in the z-axis as a voxel view as shown in Figure 5.9 (right). A voxel is a 3D extension of a 2D pixel, which may contain a vector of values indicating the features extracted from the points in it (similar to multiple color values of a pixel), or it may just contain a single bit indicating whether it has points occupied in that space or not. Many processing operations on voxels can be derived from 2D image processing in a straightforward way, e.g., 3D convolution. However, such

FIGURE 5.9 Three different representations of 3D structures.

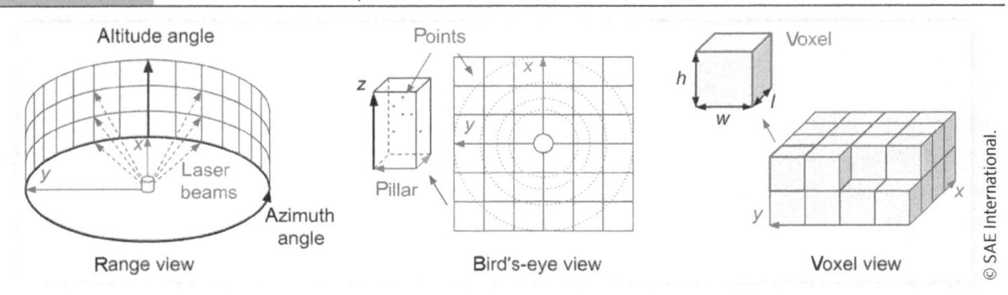

processing steps can be computationally expensive due to the increase in dimensions. Moreover, a lot of computation may be wasted since the voxels obtained from the 3D point cloud of a LIDAR scan are generally sparse, especially at a distance far away from the LIDAR. Hence, the pillar representation can be a good trade-off between the compact range image and the sparse voxels or unordered sets of points.

There is another type of 3D point cloud, usually called the 3D point cloud map of an environment, or a 3D point cloud reconstruction of an object. It can be reconstructed from multiple LIDAR scans or from camera images taken at multiple different views with precise localization of the sensors for each scanning or image and sophisticated offline postprocessing. An example of a 3D point cloud map is shown in Figure 5.10, which is computed using a visual SLAM software from images taken by a drone with GPS and Real-Time Kinematic positioning to improve localization accuracy. This type of 3D point cloud is very different from a single LIDAR scanning in density, representation, the extent of occlusion, etc., and it is usually not directly used for visual perception in CAVs.

FIGURE 5.10 A 3D point cloud map and the reconstructed HD map with texture.

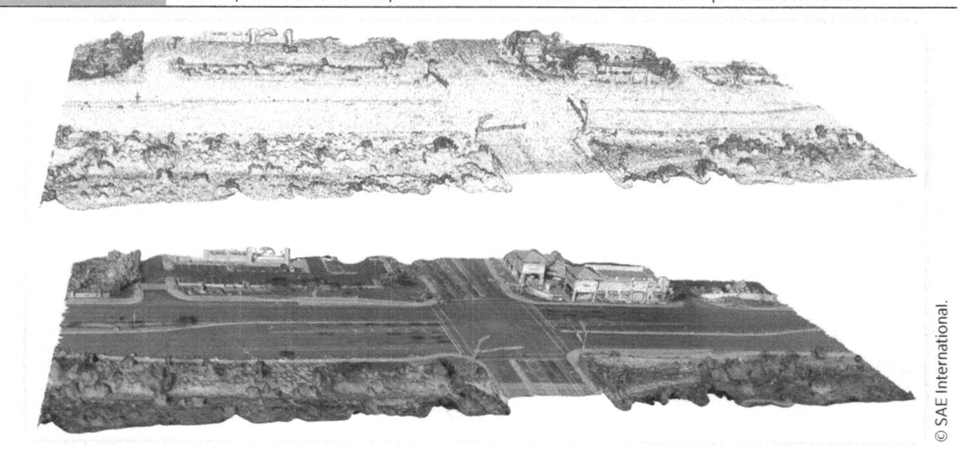

Deep Learning

People have long dreamed of computers that can understand things on the images like humans, or even computers that can think like humans. Today, AI is a broad field of study focused on using computers to do things that require human-level intelligence. It has been undergoing heavy research since the term was introduced in the 1950s. Many fascinating ideas on thinking machines in sci-fi movies, such as allowing a computer to play games like tic-tac-toe, checkers, chess, and GO, are gradually becoming reality with the advancement of AI.

Early AI systems relied on elegant mathematical formulation of the problem and efficient algorithms to solve, which means that a computer had much less knowledge about the world than a human has but the computer can implement mathematics much better than a human can. For example, in 1997, IBM made a computer called Deep Blue that defeated the world champion in chess using Monte Carlo search. Chess can be formulated with relatively simple rules in mathematics, and the search space for the moves is manageable by today's desktop computers (comparable to the "super computer" such as Deep Blue). However, around the same time, a computer struggles at many tasks that are intuitive or even trivial to humans, such as recognizing different faces, because researchers struggle at articulating the difference of faces in a mathematical way given those pixels of faces. Although AI researchers had spent a tremendous amount of effort in creating knowledge bases in formal languages, such AI systems relying on hard-coded knowledge representations were far from practical and full of inconsistency in many cases.

Meanwhile, the ability to extract patterns from data and fit complicated parametric statistical models kickstarted a new area in solving decision-related problems, and now this area is known as ML. For a long time, it generally needed a significant amount of effort from domain experts and carefully engineered pieces of information called features to constrain the complexity of the problems. For example, an ML algorithm may be able to predict whether a patient is recommended a treatment or not using logistic regression given several pieces of information about a certain collection of features of the patient, such as body temperature, age, or whether a certain symptom is present, etc. However, if the machine is given a CT scan image, it will not work. Many sophisticated image processing techniques similar to those introduced in the previous section can be applied to obtain useful features from the CT scan image. These techniques need to be carefully designed by researchers since medical image processing for diagnosis is actually a quite large and specialized discipline, where researchers can spend years to discover the underlying patterns on the images and develop a good feature extraction method. After that, domain experts with both medical and computer science backgrounds must come up with a set of rules or learning algorithms, implement them in software routines, and figure out how the data and algorithm should be connected in relation to each other to perform the specific task. As can be imagined, this is tedious and error-prone. If the data modality changes, the process must be repeated, and such a system must be redesigned from the initial feature engineering step. Clearly, the knowledge of the domain experts cannot be easily transferred to a new software.

More powerful statistical models can be used to alleviate this design difficulty. For example, online email service providers use the naive Bayes classifier to determine whether an email is a spam or not by organizing it in a "bag-of-words" feature vector. This requires less feature engineering work, and such a classifier is usually efficient in solving simple problems like answering "yes" or "no" for spam detection. However, it also requires a tremendous amount of data samples to reach reasonable performance due to the phenomenon known as the curse of dimensionality (i.e., when the dimensionality increases to a very high value, such as the number of words in a dictionary for the bag-of-words scheme, the feature space is so large that data become too sparse to reveal any pattern). Unfortunately, this is exactly the type of data that is often handled, especially in the era of multimedia content with smartphones, webcams, social media services, and all kinds of sensors that generate huge mountains of "big data" every second without human intervention. Additionally, these simple models can hardly work on more complex tasks, such as recognizing objects in an image. This poses the new challenge of understanding and extracting insights from "volumetric" data of images, speech, text content on the Internet, etc., and leads to the advent of deep learning.

Deep Learning (Goodfellow, et al., 2016) is an ML technique that automates the creation of feature extractors in terms of deep neural networks (DNNs) with layers of artificial neurons trained on large amounts of data. With DNNs, the central problem of data representations is solved by a hierarchy of complex nonlinear combinations of other simpler representations as multiple layers of neuron activation tensors, and more importantly, the way to construct these representations are learned from the raw data such as pixels in images (detailed in the coming subsections). Researchers have applied DNNs in a variety of areas including computer vision, speech recognition, machine translation, etc., and amazing results comparable to human-level performance are achieved. A type of DNN, the CNN (discussed in section "Convolutional Neural Networks"), transforms the method for many perceptual tasks on which CAVs rely, which makes automated driving close to reality. Additionally, it works not only for computer vision tasks such as image classification, object detection, and image segmentation but also for non-visual tasks such as complex and hierarchical pattern recognition, speech recognition in the real environment, and human behavior prediction for social economics. Due to its wide range of applications and significant impact on productivity, it is sometimes referred to as the "4th Industrial Revolution" (Kahn, 2016).

There were three key initial conditions that made the advancement of modern AI possible:

1. Big Data has been accumulated in nearly every business sector over the years, which provides the huge collections of data (including annotated) from which computers can learn.
2. The availability of hardware accelerators, especially GPUs, makes it possible to complete massive amounts of computation required in a practical amount of time and cost.
3. Efficient, parallel learning algorithms have been developed and implemented in the past several decades, which allows data scientists to design and train DNNs practically.

Dr. Andrew Ng, a pioneer of modern ML research, made an analogy that "AI is akin to building a rocket ship" (Garlin, 2015). In this analogy, the big data is the rocket fuel, while the learning algorithms and acceleration hardware are the rocket engine. In the field of AI, many researchers publish their cutting-edge algorithms and neural network models together with their software implementations, which are packaged up in high-level, open-source frameworks that others could reuse, i.e., they don't have to start from scratch. Meanwhile, all of the major deep learning frameworks support GPU acceleration, and cloud computing platform providers recognized the potential of deep learning to improve their own services and the business opportunity of offering GPU-accelerated deep learning platforms in the cloud. Furthermore, there are many open datasets available on the Internet, which allow researchers to investigate freely. Hence, it is now much easier to build such a "rocket ship" of AI and "launch" it to solve real-world problems. This significantly reduced the capital costs and time investment for start-ups, which motivated a huge influx of venture funding for deep learning start-ups and further accelerated the advancement of this field. Besides, large companies, governments, and other organizations—influenced by the impressive achievements of start-ups and researchers—are rapidly adopting this new technology, seeing it as a competitive advantage or threat if they fail to effectively master it.

Deep Neural Networks

The foundation of deep learning is the DNNs, also referred to as deep feedforward neural networks, or multilayer perceptions. Here, a neural network means a computation graph with multiple nodes that are analogous to neurons. The mathematical representation of such a "neuron" resembles a biological neuron, as shown in Figure 5.11 (left), where the inputs are similar to the activations of axons from other neurons, the weights are similar to the synapses on the dendrites, and the output is similar to the activation of an axon of this neuron that connects to other neurons. Hence, such a neuron can be represented as a function $h = a(\Sigma w_i x_i + b)$ given its input, and it usually "activates" nonlinearly, i.e., $a()$ is a nonlinear function, such as the rectified-linear unit (ReLU) function. Neurons can be stacked into layers such that the outputs of the previous layer are the inputs of the next layer without feedback connections or loop connections, and hence, these neurons form a feedforward neural network. For example, in Figure 5.11 (right), the neural network has an input layer with three neurons, a hidden layer with four neurons, and an output layer with two neurons. Here, each layer is fully connected with the previous layer, i.e., there is a connection for any node pair between the two layers. Typically, all neurons share the same activation function. In this way, the activation of the previous layer can be written as a vector h_{i-1}, the weights on the connections between the previous layer and the current layer can be represented as a matrix W_i, the bias of the current layer can be represented as another vector b_i, and the outputs are a vector $h_i = a(W_i h_{i-1} + b_i)$. As each neuron is a function, the whole neural network is a chain of functions of all the layers. Hence, the neural network can be denoted as a function $y = f(x, \theta)$ for succinctness, where θ indicates the weights and biases of all layers collectively, and these parameters are "learned" from the data using algorithms introduced in the next subsection. Given the input x, the neural network computation graph structure, and its parameters, i.e., θ, a program can run a forward path to compute its output y by efficiently evaluating each function of each neuron through vectorized mathematical operations.

FIGURE 5.11 An illustration of the mathematical representation of a neuron (left) and a neural network with one hidden layer (right).

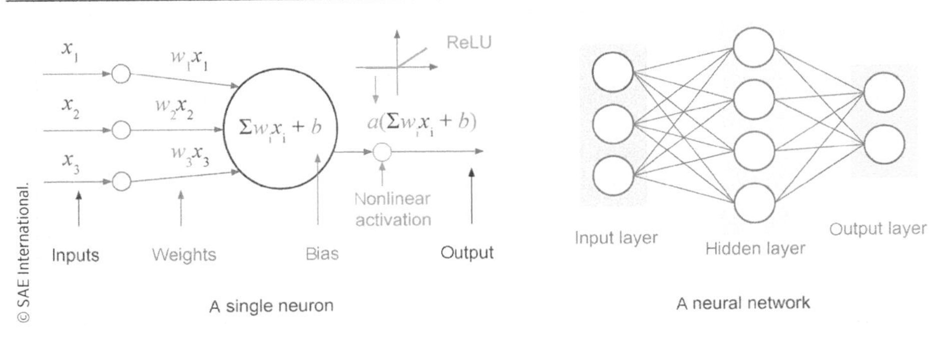

For a DNN, the term "deep" means that the neural network has multiple hidden layers. Theoretically, such a DNN can approximate any complex function $y = f(x)$ between two Euclidean spaces with enough layers and enough neurons, which is called the universal approximation theorem (Leshno, et al., 1993). It is this phenomenon property that allows computer scientists to design and train DNNs from a large amount of data to solve problems that cannot be easily solved by implementing the knowledge of a set of experts systematically in a program. For example, if x is the pixel intensities of an image and y is a label indicating whether it is a cat, dog, or other object category, such a neural network that approximates $y = f(x)$ will have the ability to recognize the content of images. However, in the real world, the architecture design of the neural network model is what makes it suitable for a particular task. For example, the best models for image classification are very different from the best models for speech recognition. There is a huge design space that researchers can explore, such as the number of layers, the number of neurons in each layer, and the connections between neurons, the activation function performed in each neuron, how the parameters are trained, etc. The Google Tensorflow Playground is a useful introduction to the DNN design space and what options are available (Smilkov & Carter, n.d.).

Training Deep Neural Networks

Given the fascinating capability of neural networks, the next step is to train these neural networks to do useful tasks. Here "training" means determining the parameters θ of the neural network. Three crucial components are needed to successfully train a neural network for a specific task.

First, a training dataset consists of a collection of representative data examples must be assembled. The training can be accomplished with either unsupervised or supervised learning. In unsupervised learning, data are unlabeled and so the DNN has to find underlying structure in the data without assistance; this technique is often used to pre-process data by clustering like data groups and compressing the data without losing useful information. In supervised learning, data are labeled (or annotated), sometimes by a domain expert, sometimes by cheap labor (i.e., Mechanical Turk or offshored entity), sometimes by the user (i.e., Facebook tagging of friends). For example, if this neural network is asked to classify images with cats or not, usually several thousand images of cats and another several thousand images without cats need to be collected. If supervised learning is employed, each image must be associated with a label indicating whether it belongs to the cat class or not. To ensure the dataset is representative of all the pictures of cats that exist in the world, the dataset must include a wide range of species, poses, and environments in which cats may be observed. If the dataset is too small or not representative, the trained neural network would be biased, and it may only be able to recognize cats in images that are similar to the images in the training dataset.

Second, a neural network architecture with a loss function must be designed. Taking the image classification of cats and dogs as an example, a neural network may be designed with multiple layers. The input layer may have a dimension comparable to a 100×100 pixel image with three color channels, i.e., 30,000 neurons. Then a few hidden layers may be stacked, each with a few thousand neurons and fully connected with the previous layer. After that, a single neuron may be placed in the output layer that predicts whether an input image subject is a cat or not. This output neuron will use a sigmoid function as its activation function, and its output is denoted as y', which is some number between 0 and 1 indicating the probability or confidence of the prediction that this image subject is a cat. For each input image, assume that the ground truth is available and labeled as y, i.e., if y is 1, the image subject is a cat, and if y is 0, it is not. Finally, for any single input image, a loss function can be defined as $L(y, y') = -y \log(y') - (1 - y)\log(1 - y')$. The value of this loss function is designed to be a positive number. It is large when the prediction is far from the ground truth, and it should be close to 0 if the prediction is close to the ground truth, and hence, it is called a loss function or a cost function. If a collection of N images are given as a mini-batch, the loss function can be defined collectively as $L = -\frac{1}{N}\sum_{i=1}^{N} y_i \log(y'_i) + (1 - y_i)\log(1 - y'_i)$. Such a loss function is called a binary cross-entropy loss, which is widely used in binary classification ML models. It can be extended to accommodate multiple categories instead of just two shown in this example.

Third, a training algorithm must be implemented. The training target is minimizing the loss function by adjusting the neural network parameters such that the predicted label should be close to the ground truth label for all the images in the training dataset. Different from optimizing a simple mathematical function of polynomials, the existence of the complex neural network architecture, the large number of parameters between each layer, and the nonlinear activation all make this minimization step difficult. Currently, the standard practice is gradient-based learning with backpropagation. Given any loss function L, gradient-based learning means calculate the first-order partial derivative of the loss function with respect to the parameters, i.e., the gradient, and iteratively adjust the parameters until they converge. Mathematically, this procedure can be represented as $\theta^{k+1} = \theta^k - \lambda\frac{\partial L}{\partial \theta}$, where k means the kth iteration, and λ is the learning rate that controls how much the parameter should be updated in each iteration. The initial value of the parameter is typically set randomly. To implement this, the training algorithm will select a mini-batch of input data and run the forward path of the neural network to calculate a loss value. Then we can calculate the gradient of the loss with respect to each neuron of the last layer and use the chain rule of partial derivative calculation to propagate the gradient to the layer before, until the first layer. Mathematically:

$$\frac{\partial L}{\partial \theta} = \frac{\partial L}{\partial h}\frac{\partial h}{\partial \theta} = \frac{\partial L}{\partial h}\frac{\partial h}{\partial g}\frac{\partial g}{\partial \theta} = \cdots \tag{1}$$

where $h()$ is the function representing a specific neuron in the last layer, $g()$ is another neuron in a layer previous to the last layer, and so on. This allows for a calculation of the gradient of the loss function with respect to each weight and bias parameter and applies the gradient-based optimization on each parameter. As the neuron activation computing is done layer by layer from the input layer to the output layer in the forward path, this gradient of loss function computing is done in a similar way layer by layer but in the reverse direction, and hence, it gets the name "backpropagation." In this way, one learning step involves a mini-batch of input data, a forward path neuron activation computing, a backward path gradient computing, and one iteration of parameter updating.

Eventually, after many such steps with randomly selected mini-batches of data, the neural network will hopefully converge and make correct predictions for most of the training data. Typically, the neural network is said to have been trained for one epoch if the training procedure has used enough mini-batches of data comparable to the size of the whole training dataset. This means that after processing the entire training dataset once, the neural network should generally have enough experience to choose the correct answer a little more than half of the time (i.e., better than a random coin toss to make a prediction), but it will usually require additional tens of epochs to achieve higher levels of accuracy. Sometimes the loss will reach a plateau, i.e., it will not decrease with more rounds of training, but the neural network still cannot make good predictions. In this case, a change in the neural network architecture design may need to be considered, for example, using CNNs introduced in the next subsection instead of fully connected neural networks to gain better performance. In some other cases, the training may just fail, i.e., the loss stays unchanged, or bounces, or becomes NaN (i.e., Not-a-Number in the definition of floating point number representations, usually generated through dividing by zero or other invalid mathematical operations). This may be caused by the gradients vanishing to zero or exploding to infinite after too many steps in the chain calculation. Hence, changing the learning rate, the mini-batch size, the design of the loss function, or even rethinking the task itself must all be considered. As the parameters of a neural network are completely calculated from the data, the quality of the data, the quality of the label, and the relevance of the task matter a lot. Otherwise, a neural network will simply show the phenomenon of "garbage in and garbage out."

Convolutional Neural Networks

As a special type of neural network, the CNN has revolutionized visual recognition research in the past decade. Different from the traditional handcrafted features, a CNN uses multiple specialized convolutional layers and pooling layers to efficiently capture sophisticated hierarchies describing the raw data, which has shown superior performance on standard object recognition benchmarks while utilizing minimal domain knowledge. This special convolutional and pooling architecture is inspired by neuroscience discoveries on the visual cortex of primates, which shows a similar structure with biological neurons.

In Figure 5.12 (left), a simple CNN is shown with an input layer of six neurons, a convolutional layer of four neurons, a polling layer of two neurons, and two fully connected layers. The convolutional layer is similar to a normal layer, but it is only locally connected (e.g., the neuron c_1 is only connected to input x_1, x_2, and x_3), and the connections share weights among the neurons (e.g., all the connections with the same color share the same weights). This is equivalent to applying a convolution operation on the input or the activations

FIGURE 5.12 An illustration of a CNN.

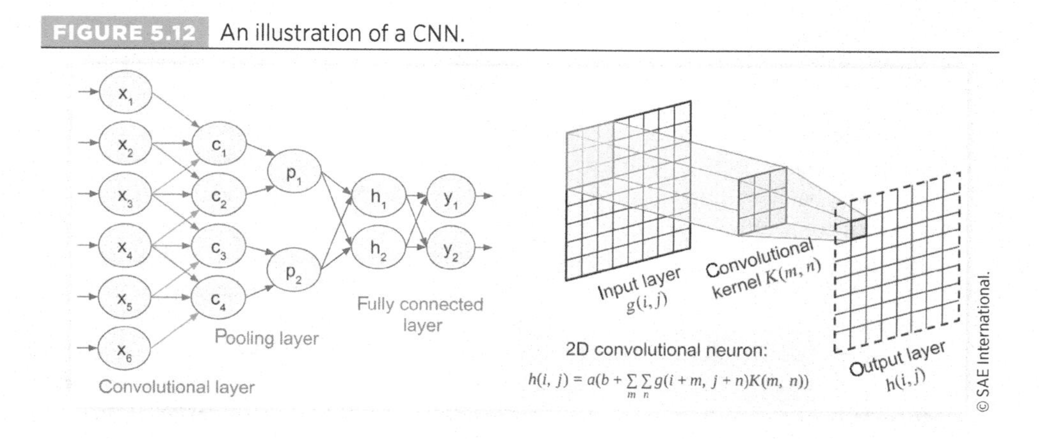

of the previous layer, while the convolutional kernel contains the trainable weights, as shown in Figure 5.12 (right). As mentioned previously in section "Image Processing," different values in the convolutional kernel can have drastically different results, and such results can be considered as a special type of image feature defined by this convolutional kernel. In this figure, only one kernel is shown while, in practice, multiple different kernels can be applied on the same input to generate outputs like images of multiple channels, where each channel means a different type of feature. Hence, these convolutional kernels are also called image filters. As a kernel scans over an input, the extracted features generate another image, which can be processed by another kernel in another layer. Besides, the pooling layer will just select the strongest activation from the input neurons, e.g., in Figure 5.12 (left), the neuron activation of p_1 is just the maximum value of its inputs from c_1 and c_2. This works as a downsampling process that summarizes and abstracts the features. In a CNN, multiple convolutional layers and pooling layers can be stacked to aggregate low-level features to high-level features. Meanwhile, these kernels are learned from data so that the neural network can be trained to automatically extract useful features to the final task corresponding to the loss function.

For example, if the objective is to recognize images with cats, it is desirable to locate high-level features such as eyes, noses, ears, limbs, and furry body, which define a cat. Each type of the high-level feature of a component of a cat can be further described as a collection of shapes and texture patterns, which can be further broken down as a set of low-level features such as corners, line segments, and color patches. Although a CNN does not work exactly in this way, this analogy can basically explain the underlying principle. Another example is shown in Figure 5.13, which is a well-known network called LeNet (LeCun, et al., 1998) designed by Dr. Yann LeCun, and probably it is the first successful application of CNNs. This neural network is designed to recognize handwritten digits from 0 to 9. The input is a grayscale image of 32×32 pixels. The neural network consists of two sets of convolutional-pooling layers with the same kernel dimensions, but different numbers of filters. After that, two fully connected layers are applied to classify the aggregate features vectors into ten categories. The network is trained on the MNIST dataset with over 50,000 images, and some sample images are shown on the left side of Figure 5.13. Once trained, it can achieve over 99% of accuracy on a separate set of about 20,000 similar images.

FIGURE 5.13 The LeNet for handwritten digits recognition.

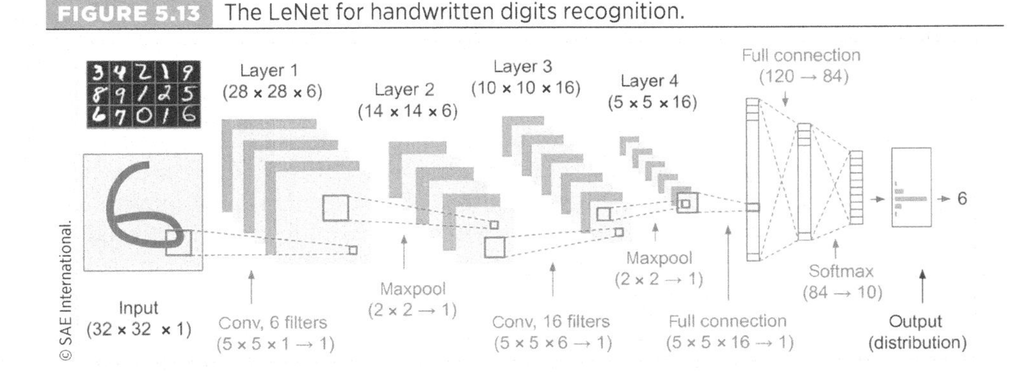

Perception Tasks for CAV

Equipped with the basics of sensors and data representation, now four essential perception tasks for CAV are explored, including (1) object detection, (2) object tracking, (3) segmentation, and (4) 3D depth estimation from images. For each of the tasks, the problem is formulated under the context of CAV, and the challenges, as well as solutions, are discussed.

Object Detection

Just as a human driver needs to be aware of other vehicles, pedestrians, and traffic signs, detecting these objects in a reliable and prompt way is crucial to CAVs, as shown in Figure 5.14. Depending on the sensor modality, object detection can be done on a 2D image or a 3D point cloud. For a 2D image $I(x, y)$, the output of an object detection module is a list of objects of known categories. The location of each object is represented as a 2D bounding box (x_1, y_1, x_2, y_2) on the image, where (x_1, y_1) and (x_2, y_2) is the top-left corner and bottom-right corner of the box, respectively. Additionally, each object has a predicted class label and a prediction confidence score, like the image classification task discussed in section "Training Deep Neural Networks."

FIGURE 5.14 An example of 2D object detection on a 2D image.

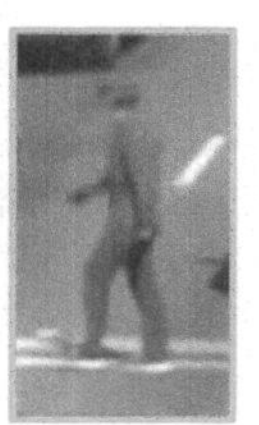

Although object detection seems instinctive for humans, the definitions of "objectness" and visual saliency are not always clear for a computer with a camera, especially with cluttered image background seen from a CAV driving in urban areas. Generally, this problem is solved by a neural network trained on a large dataset of images with human annotations. For example, in Figure 5.15 an overview of the Faster RCNN object detector (Ren, et al., 2015) is shown, where RCNN means "Region-based Convolutional Neural Network." It contains three main components. First, the input image is processed by a backbone network to generate a hierarchy of feature activations. This network contains only convolutional layers, and hence, it can be applied to images with any size or aspect ratio. Second, the region proposal network scans each level of the generated feature hierarchy in a sliding window fashion to generate candidates of regions that may have an object. Third, each region candidate is interpolated to a fixed size and further processed by a head neural network with a set of fully connected layers to predict its object category and refine the box shape. Finally, these bounding boxes and class labels are packed into data structures and returned as the results.

FIGURE 5.15 An illustration of the architecture of the Faster RCNN object detector.

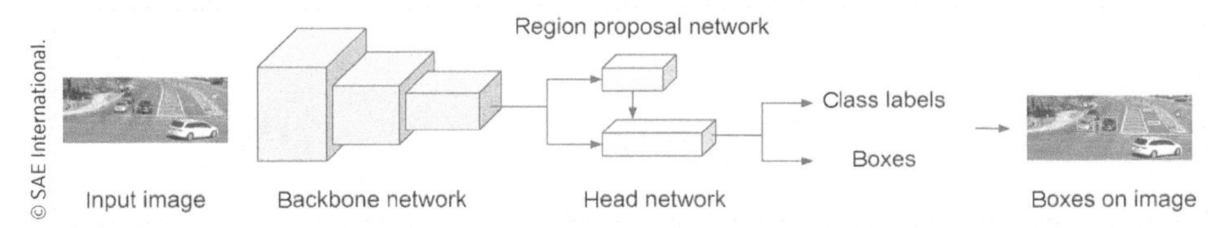

For a CAV, knowing the objects on 2D images is not enough. The 3D shape, location, and orientation of these objects are desired to prevent crashing into them. This problem is 3D pose estimation, and it is usually solved by an algorithm or another neural network connected immediately after the object detector. For objects with a rigid "boxy" shape such as a vehicle, a 3D bounding box with orientation is needed, while for pedestrians and cyclists, either a bounding cylinder (i.e., a center point on the ground, a radius, and a height) or an articulated model containing positions of each joint of the person is desired, as shown in Figure 5.16. One possible solution is detecting important key points on the 2D image and then recovering the 3D pose using the camera parameters, the prior knowledge of the object, and some geometric constraints. For example, in Figure 5.16 (left), a neural network can be used to detect the corner points of the 3D bounding boxes on the image in a way similar to computing the 2D bounding boxes in object detection. After that, the 3D pose of the vehicle can be calculated by solving the well-studied perspective-n-points problem using the camera projection, the prior knowledge of vehicle dimension, and the fact that all four corner points on the bottom surface of the 3D bounding box are on the ground.

FIGURE 5.16 Examples of 3D pose estimation.

Besides the indirect way of detecting the objects and estimating their poses from an image, a CAV can also directly detect the 3D bounding boxes of objects on a point cloud obtained from a real-time LIDAR scan, as shown in Figure 5.17. Typically, this is also achieved by a neural network. For example, the PointPillars detector (Lang, et al., 2019) runs three neural networks. The first one extracts features on the pillars of the point cloud and generates a pseudo image of feature activations in the bird's-eye view. The second one is a 2D CNN that further computes a hierarchy of features on this pseudo image, like the backbone network in 2D object detection. The third one is another neural network called the head network that does the actual 3D object detection from the feature activations.

FIGURE 5.17 An example of the 3D point cloud from three real-time LIDAR scans obtained at the same time but in three different places. The LIDARs 3D point cloud data are from the company Luminar.

Tracking

Unlike an object detector that processes only one image, an object tracker needs to estimate the trajectory and velocity of multiple objects given the continuous measurement of the camera or the LIDAR. This is very important for a CAV to predict the future behavior of other traffic participants and avoid collisions. There are a few major challenges, including the complexity of motion, occlusions, change of object appearances under different view angles, objects with similar appearances interacting with each other (e.g., pedestrians), and cluttered backgrounds (e.g., urban streets).

One straightforward solution is tracking-by-detection, which means running the detector independently on each image or LIDAR scan and associating the detected objects on the current sensor readings to those objects detected and tracked previously. In this way, tracking is essentially reduced to an object association problem. Formally, given a list of objects already tracked and a list of newly detected objects as two disjoint sets of nodes in a bipartite graph, the objective is to compute a set of edges between the two sets that best match the nodes by minimizing some overall cost one these edges. Typically, a real object in the 3D world does not move significantly during the interval of two consecutive camera images or LIDAR scans, and the extent of bounding box overlap can be a good cost metric for an edge connecting the two nodes. Such edge cost can also be computed from various attributes of a pair of tracked objects and a newly detected object such as the consistency in color and shape, displacement, etc.

Besides the simple object detection and association method, a tracker can also use a Bayesian inference framework, typically some variant of Kalman filter or some nonparametric Bayes filter (e.g., particle filter), to iteratively estimate the position, velocity, and other states of the object in the tracking process. Note that these states are represented by random variables, and such a filter calculates the probability distribution of these variables in a parametric or nonparametric form upon the arrival of each "observation," e.g., one iteration for each new image taken by the camera. In each iteration, the states of the object are first predicted given their current probability distribution (i.e., results of the previous iteration) using the kinematics or dynamics property of the object. Then an association algorithm assigns newly detected objects to existing objects using the predicted results, or a joint detection and tracking step can run here to detect objects on the new image using the predicted results and other features such as the pixel-level motion discussed below or some feature embeddings from a DNN. After that, the states are updated with the new observations, i.e., the posterior probability distributions of the state variables are calculated using the Bayes rule. This framework has been widely used in the past for its simplicity and good performance on rigid-body objects with well-studied dynamics, e.g., vehicles. However, it has problems when there are wrong associations or missing observations due to occlusions. With the advancement of DNNs, there are also end-to-end joint detection and tracking methods that utilize RNN or both RNN and CNN to solve the tracking problem. However, the lack of large amounts of annotated data (especially annotation on every video frame) makes such solutions less practical.

Tracking can be done not only at the object level but also at the pixel level or 3D point level. On a pair of 2D images taken by a camera consecutively, the optical flow is defined as the 2D motion vectors of each pixel between them. Similarly, scene flow is the motion between two point clouds from two consecutive LIDAR scans, which is a generalization of optical flow in 3D. However, there are generally no point-to-point correspondences because it is not the points that move. In fact, the points are generated by the laser beams that sample moving objects. Hence, scene flow is usually represented by a collection of 3D vectors indicating the motion of each point on the first point cloud.

To compute the optical flow, typically an assumption is made that a pixel does not change its brightness but "moves" to a new place with a displacement vector, i.e., $I(x, y, t) = I'(x + \Delta x, y + \Delta y, t + \Delta t)$. If the right-hand side of this equation is expanded using the Taylor series, keeping only the first-order term, and we cancel out the $I(x, y, t)$ on the left-hand side and divide every term with Δt, we have $\frac{\partial I}{\partial x}\frac{\Delta x}{\Delta t} + \frac{\partial I}{\partial y}\frac{\Delta y}{\Delta t} + \frac{\partial I}{\partial t} = 0$, or $I_x V_x + I_y V_y + I_t = 0$, where I_x and I_y are the gradient of the pixel value in x and y directions, (V_x, V_y) is the optical flow vector that needs to be solved at that pixel, and I_t is the derivative of the brightness of that pixel regarding the time. Clearly, given a single pixel, i.e., one equation with two unknowns, (V_x, V_y) cannot be solved. As a result, all solutions require some additional assumptions globally on the image or semi-globally on a local patch. For example, the Lucas-Kanade method (Lucas & Kanade, 1981) solves this by assuming that a local image patch of $m \times m$ pixels has the same optical flow vector. However, this cannot work on image patches with little or no brightness change, e.g., a featureless surface with the same color, or no change in one direction, e.g., on a line. Hence, it is used to track the motion of a set of detected corner points, which are usually sparse, e.g., in the KLT tracker (Tomasi & Kanade, 1991), as shown in Figure 5.18 (right). Here the sparse optical flow vectors along the object moving direction are shown in blue and others are in red. Other, more advanced methods, as well as deep learning models, are also invented to solve dense optical flow on the whole image. However, accurate optical flow computing still faces the same challenges as object tracking and it is generally more computationally expensive.

FIGURE 5.18 An example of object tracking and sparse optical flow (the short blue line segments).

One video frame

Vehicle tracking

Sparse optical flow

Segmentation

Distinct from detection or tracking at the object level, image segmentation means partitioning the image into regions and assigning a label to each region from a set of predefined categories. There are three types of segmentation, as shown in Figure 5.19. The first one is semantic segmentation, which is essentially a classification of each individual pixel. It generates a mask of the image with the same width and height, and each pixel of the mask is the category label of that pixel on the original image. For example, in Figure 5.19 (top-right), each pixel of the image is assigned with a semantic label such as person, vehicle, traffic light, traffic sign, road, curb area, etc. In this case, pixels on two different vehicles are assigned with the same label. The second one is instance segmentation, which is simultaneously detecting objects of an image and generating a semantic segmentation mask for each of those objects. The output is similar to an object detector with a list of object bounding boxes and a list of class labels, plus a list of masks. In this case, a mask contains only pixels on a specific object, and pixels on two different vehicles are assigned with the same semantic label, but they are on two different masks. Those pixels not on any object are considered "background." The third one is panoptic segmentation, which is a combination of semantic segmentation and instance segmentation. In this case, every pixel of the image is assigned with a semantic label and an instance ID. For example, in Figure 5.19 (bottom-right), those pixels on two different vehicles have the same semantic label, but different instance IDs. The panoptic segmentation contains masks of both countable objects (i.e., vehicles, pedestrians, traffic signs) and uncountable amorphous regions of the same texture (i.e., road, sky, curb area). For a CAV, this can provide a coherent scene understanding of what is seen.

FIGURE 5.19 Examples of three types of image segmentation.

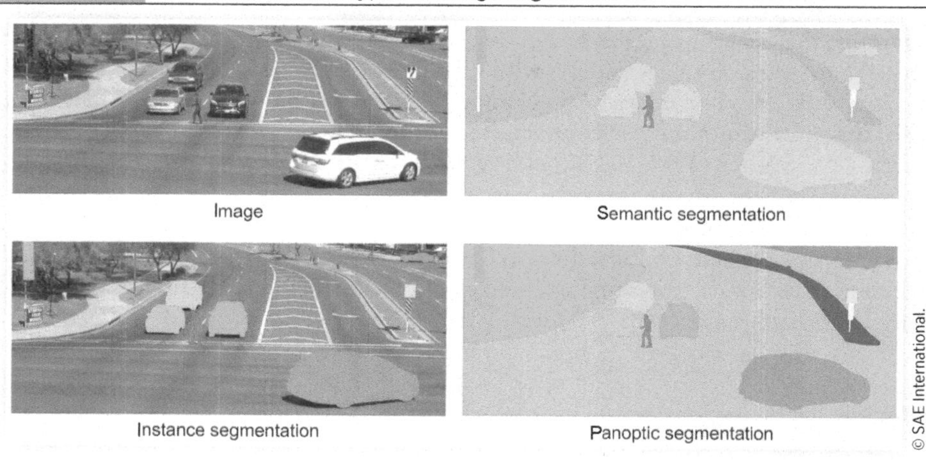

Inspired by the success of neural networks on image classification and object detection, there are many deep learning solutions that use CNNs to generate high-quality segmentation results. However, a traditional CNN for object-level tasks will combine the features from multiple scales of the image to high-level semantic meanings, while segmentation requires both high-level semantics and pixel-level boundaries. A popular method to tackle this issue in semantic segmentation is using an encoder-decoder network structure. For example, in SegNet (Badrinarayanan, et al., 2017), the image features are extracted and downsampled by a fully convolutional encoder network with pooling layers similar to a network for image classification without the fully connected layers. After that, the generated feature maps pass through a decoder network with an identical architecture but in a top-down fashion, where the pooling layers are replaced by upsampling layers using the pooling indices obtained in the encoder. Eventually, the decoder generates feature maps in the same resolution as the original image, and a pixel-wise softmax layer is used to obtain the class label for each pixel. This encoder-decoder idea can also be used for instance segmentation and panoptic segmentation. For example, in Mask RCNN (He, et al., 2017), which has a similar network architecture as that in Figure 5.15, the encoder is the backbone network, and the decoder is in a branch of the head network that generates binary masks for each object. These methods have achieved impressive results, especially in the scenarios for automated vehicles, but there are still a few challenges to address, such as accurate segmentation of small objects with a relatively far distance and the need of huge computation (most networks for pixel-level segmentation tasks cannot run at real-time on a high-end workstation GPU with a reasonable image resolution, let alone on embedded devices on CAVs).

3D Depth Estimation

Humans can easily perceive the rough distance to an object without relying on sensors that actively emit energy to the environment such as LIDAR. This is achieved partially through the difference of the visual information between the left eye and the right eye, and partially through the prior knowledge of the object size in 3D and its apparent size. For a CAV, due to the high price tag and reliability issues of a LIDAR, it is desired to obtain the distances to the surrounding obstacles using one or two cameras, i.e., estimating the 3D depth of each pixel (this was discussed in Chapter 4).

Given two images from two cameras observing the same 3D point, the 3D depth of this point can be obtained through triangulation, as shown in Figure 5.20 (left). Assume we know the centers of the two cameras for the two images, i.e., C and C'; the relative rotation and translation from the second camera to the first camera, i.e., (R, t); and the intrinsic matrices of the two cameras, i.e., K and K'. This essentially means the projection matrices P and P' of the two cameras are known. A 3D point X will be projected to 2D points x and x' on the two images. It is clear that X, C, and C' form a plane that intersects with the two image planes at two lines l and l'. This property is called the epipolar constraint, and the two lines are called epipolar lines. Using this property, given any point x on the first image, we can derive the line l' from the camera parameters and search along this line to find its corresponding point x' by assuming they have the same appearance (similar to that in optical flow computing). After that, the location of the 3D point X relative to the first camera can be obtained by solving the equations $x = PX$ and $x' = P'X$ together.

FIGURE 5.20 Generic two-view depth triangulation (left) and stereo depth estimation (right).

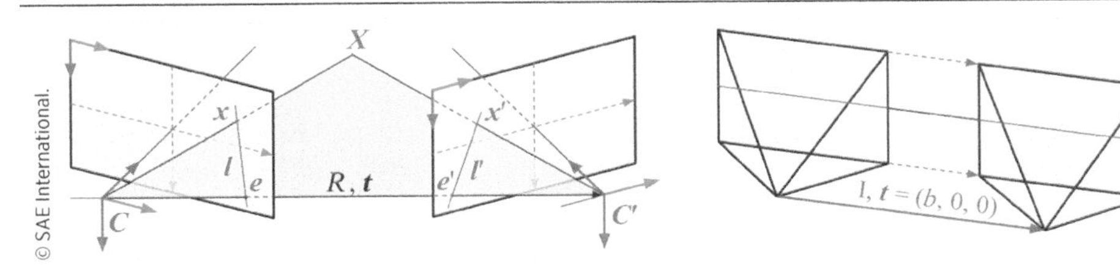

To make the point search easier, typically two identical cameras are specially oriented in tandem such that the images are in "row correspondence," i.e., both cameras have identical intrinsic matrix K, both images have the same resolution, x and x' are in the same row, and there is only a translation $(b, 0, 0)$ between C and C', as shown in Figure 5.20 (right). Such a camera configuration is usually called stereo-vision. In this simplified case, the Z coordinate of the 3D point X, i.e., the depth, can be calculated as $Z = fs_x b/(x - x')$, where fs_x is the first element on the diagonal of K; b is the distance between the two cameras, typically called the baseline distance; and $(x - x')$ is the difference between the x coordinates of the two corresponding points, typically called the disparity. In Figure 5.21, an example of such a stereo-vision camera configuration and an example result of stereo-vision depth triangulation are shown. Note that the point correspondence search may fail, especially when the images contain featureless areas, e.g., the white walls and ceiling in Figure 5.21. Moreover, the depth resolution degrades with the distances, especially when the 3D point is far away and the disparity $(x - x')$ is below one pixel. Furthermore, the choice of the baseline is also important to achieve good depth resolution at some desired distance. Hence, the raw depth results may have errors or invalid values (i.e., "holes"), which require some postprocessing steps (which is usually called depth completion). Recently, deep learning has been applied for depth completion or even direct disparity generation, which has achieved impressive results. Besides, a DNN can even predict the depth from a single image with a reasonably good accuracy given enough training data.

FIGURE 5.21 An example of stereo depth estimation.

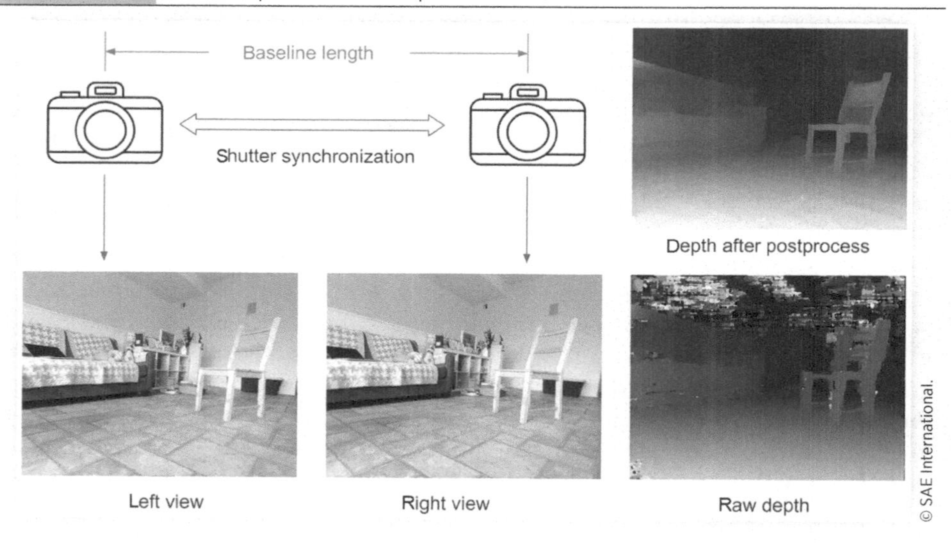

Perception System Development for CAV

The perception system of a CAV usually contains several interdependent modules to process the information from the visual sensors for both low-level and high-level computer vision tasks. The inputs to the perception system are HD maps, the location of the vehicle, and the sensor readings, e.g., images from cameras or 3D point clouds from LIDARs. The sensors are calibrated relative to the vehicle and each other in the factory. Given the sensor calibration and configuration parameters, the outputs of these processing modules are fused together to provide a complete understanding of the surrounding traffic scene, including the detected traffic lights, lane markers, and the 3D bounding boxes, as well as current motion states of vehicles, pedestrians, and other traffic participants. Typically, there is also a prediction module that uses the current observation to generate expected future trajectories of other vehicles and traffic participants. This modularized design enables development of each sub-system in parallel and evaluation of the performance in a systematic and interpretable way. However, there is also recent progress on end-to-end learning for automated driving without relying on any intermediate perception representations.

As the perception system of a CAV relies on trained models, especially DNNs, large-scale datasets have been the key in the development and benchmarking. Many open datasets from academia for general image understanding (e.g., ImageNet, COCO, etc.) and specially for perception in automated driving (e.g., KITTI, Berkley DeepDrive, etc.) have made significant contributions in accelerating the research field. The industry has also invested a significant amount of resources in the dataset construction, and every major CAV company maintains its own dataset assets. Parts of these kinds of datasets have been made publicly available by these companies for researchers in academia, such as ApolloScape, NuScenes, Argoverse, and Waymo Open Dataset. Most of the datasets are used to evaluate the model-level performance of a specific task discussed in section "Perception Tasks for CAV." However, evaluation of the operational performance and robustness of the perception system of CAVs is still an open question, which may be answered by large-scale public road testing currently carried by the industry. Since the DNNs used in many perception modules are black boxes even from the developers' point of view, the robustness of perception results in rare cases is especially important. On the other hand, with the fast advancement of sensor technologies and processing algorithms, the capability of the perception system of a CAV in the future may be different from that on current vehicles. In the remainder of this section, we briefly discuss three example perception systems for CAVs as case studies.

Case Study: Google/Waymo CAV

The development of Google/Waymo's CAV began in 2009, motivated partially by its effort in new mapping and street view technology (Waymo was spun off from Google in 2016). Many of the early team members are from the Stanford team that won the 2005 DARPA Grand Challenge on automated driving. At that time, efficient visual perception algorithms were still challenging, considering the computation power of CPUs and DNNs for visual understanding tasks were not well developed yet. In the DARPA Grand Challenge vehicle, a Velodyne 64-line LIDAR was used as the main sensor. The LIDAR could generate point clouds with an impressive amount of detail with multiple scans every second. The team collected a huge amount of data and created an HD point cloud map of the San Francisco Bay Area, including detailed geometric features in centimeter-level resolution, dense LIDAR reflection intensity, and lane-level road semantics. With this map, the CAV could localize itself in real time by matching the LIDAR reflection intensity features from a scan and this map. Objects in the environment can be detected and classified. Additionally, a dash camera was used to recognize the traffic light signals.

As most of the technology details of how Waymo's CAVs perceive the environment are kept unknown for the general public, the big picture can be found through a keynote speech at *the Annual IEEE International Conference on Intelligent Robots and Systems* in 2011 (Thrun & Urmson, 2011). This perception system has seen minor upgrades since then with more vehicles added into the test fleet, and it has been tested for several millions of miles of automated driving in multiple cities in the coming years. In 2017, a major redesign occurred, and the company equipped their CAVs with a suite of sensor hardware developed in-house, which drastically reduced the cost. The new design features multiple cameras, RADAR units, and five LIDAR units (one on the top, and four smaller LIDAR units on the four sides of the vehicle). Due to the 3D sensing capability and consistent performance under different lighting and weather conditions, LIDAR is chosen as the main sensor. Illustrations of the 360° LIDAR sensing results, the actual vehicle after the 2017 redesign, and an illustration of the sensors are shown in Figure 5.22.

FIGURE 5.22 An illustration of a Google self-driving car after 2017.

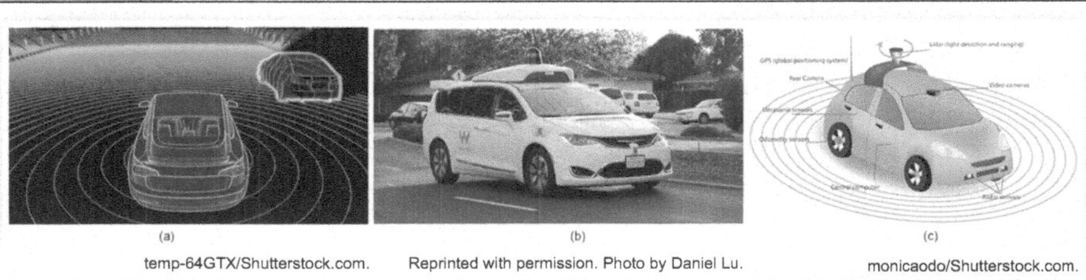

(a) temp-64GTX/Shutterstock.com. (b) Reprinted with permission. Photo by Daniel Lu. (c) monicaodo/Shutterstock.com.

Case Study: Tesla Autopilot

Distinct from Waymo CAVs that aim to achieve Level 4 autonomy directly, Tesla Autopilot is a combination of active safety and Level 2 automation features that could enable the vehicle to gradually evolve from Level 2 to Level 3 and eventually achieve Level 4 with its Full Self-Driving (FSD) package. The first version was released in 2014. In the early stages, Tesla vehicles relied on Mobileye for their active safety capabilities. Later, a mix of Nvidia computing platforms and sensors from various suppliers were used. Currently, a Tesla vehicle with Autopilot has a set of Level 2 features, including lane-keeping, ACC, self-parking, etc. However, the driver is still required to take responsibility, i.e., keep hands on the steering wheel, keep eyes on the road, and engage as needed; the driver is responsible for part of the OEDR. In late 2020, Tesla unveiled a beta version of its FSD hardware and software that can demonstrate impressive behavioral and perception capabilities using a combination of camera, RADAR, and ultrasonic sensors. For example, a Tesla Model 3 with FSD can recognize the lane markers, traffic signs, and other vehicles (in terms of 3D bounding boxes), as shown in Figure 5.23. However, users and experts in the field still have doubts on whether the current Tesla vehicle configuration has all the necessary hardware and software for FSD, especially without using any LIDAR and HD maps. Tesla has claimed that its vehicle can use the RADAR to obtain a point cloud similar to a LIDAR to handle environments with low visibility, and it also argues that relying on HD maps to drive makes it overall "brittle" to the changes in the physical environment. More recently, Tesla has moved away from using RADAR at all, and is focusing on camera-only computer vision. This makes Tesla an outlier in the major CAV developer community, but only time will tell which approach will become dominant.

FIGURE 5.23 A screenshot of the perception results of a Tesla Model 3 with FSD on the dashboard of the vehicle.

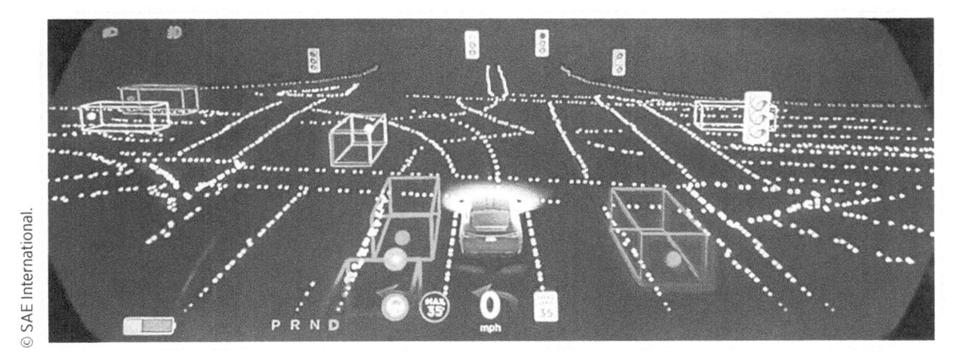

In the most recent version, a typical Tesla vehicle, e.g., a Model 3, has eight surround cameras to cover a 360° FOV around the vehicle. Among them, two forward-looking side cameras are placed on the center pillar, two backward-looking side cameras are placed between the front door and the front wheel, one rear-view camera is at the back of the vehicle close to the trunk handle, and three forward (wide angle, main, and narrow field) cameras are inside the vehicle at the top of the front windshield close to the rear-view mirror. It is also equipped with two Tesla-designed FSD chips with computation power comparable to workstation GPUs to run large neural networks for both perception and decision-making. These neural networks are trained on data collected from a large number of human drivers and tuned through a "shadow mode" on Tesla vehicles on the road to mimic the behavior of a person. As the Tesla FSD function is still under testing, with both exciting positive opinions and criticism, its actual operational performance is still yet to be determined.

Case Study: CAROM

CAROM (Lu, et al., 2021) stands for "CARs On the Map," which is a system for vehicle localization and traffic scene reconstruction using monocular cameras on road infrastructures developed by the authors of this book together with the Institute of Automated Mobility (IAM) and Maricopa County Department of Transportation (MCDOT) in Arizona, USA. The motivation is to analyze naturalistic driving behavior, assess the operational safety of automated vehicles, conduct fine-grained traffic statistics, etc., all from the "third-person view" through the road infrastructure, especially through traffic monitoring cameras that are already deployed on the traffic light poles managed by local DOTs such as MCDOT. Also, these cameras are considered as an essential component of the intelligent transportation system in the future because they are typically placed at strategic locations for road traffic monitoring, and their sensing and communication capability can be helpful for both the automated vehicles on the road and the regulators who manage the infrastructure.

The vehicle tracking and localization pipeline of the CAROM system contains multiple stages, as shown in Figure 5.24. For each incoming camera image, object detection and instance segmentation are implemented to obtain the mask of each object. Then the sparse optical flows are used to track the vehicles across frames. After that, the camera parameters are used to calculate the three vanishing points for the XYZ axes of each vehicle and draw tangent lines to the contour of the vehicle mask to compute the 3D bounding boxes. The camera is modeled as a pinhole camera and calibrated using a set of point correspondences

FIGURE 5.24 The CAROM pipeline on vehicle tracking and localization.

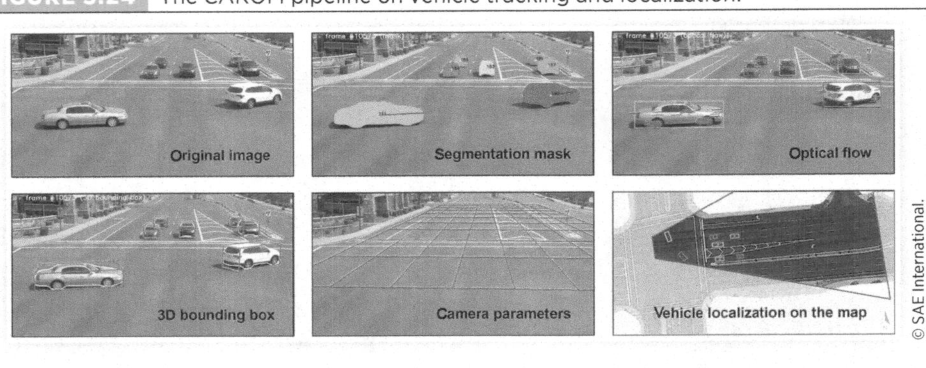

manually labeled on the image and the map. Since the camera does not move after installation, this calibration only needs to be done once for each camera. Assuming the ground area covered by the camera is flat, any point on the image can be transformed to a point on the map and vice versa. With this transformation between the image and the map as well as the tracking and pose estimation results on the image, the accurate location and speed of each vehicle can be determined on the map as well as in the 3D world (because the map is to-scale).

The tracking pipeline was applied to the traffic monitoring videos obtained from four cameras pointing to the four directions of an intersection in the city of Anthem around the Phoenix metropolitan area in Arizona (Altekar, et al., 2021). The cameras are mounted on the traffic light infrastructure and their useful depth of view is roughly 120 m. These videos allow for an observation of the same vehicle traveling through the intersection from multiple viewing angles. Given a sequence of vehicle locations, 3D bounding boxes, and segmentation masks from multiple images, the vehicle's visual hull is computed using the shape-from-silhouette method, and then the reconstructed 3D shape of the vehicles are refined through a model fitting step (i.e., using the 3D models of the real vehicles to correct the reconstructed models so that they resemble real vehicles more closely). The vehicle tracking, localization, and traffic scene reconstruction results have been evaluated quantitatively and qualitatively. To obtain the ground truth of vehicle locations, a drone was flown to capture the vehicle trajectories from above, and a test vehicle equipped with a high-precision differential GPS to record its motion states was also driven through the intersection. Based on these tests, more than 90% of the vehicles are correctly tracked. Errors usually happen on those vehicles that are partially or totally occluded. Also the localization error is approximately 0.8 m and 1.7 m on average within the range of 50 m and 120 m from the cameras, respectively. The tracking results are compact data structures, which are much easier to transmit, archive, index, and analyze compared to the raw videos. Moreover, these results do not have the privacy issues of the videos, and they can be shared and "replayed" in 2D or 3D so as to allow them to be accessed by other people not directly affiliated by the DOT, such as analysts in third-party companies and scholars in academia. In the future, the aim is to continue development of the CAROM algorithm to decrease measurement uncertainty and then to widely deploy CAROM algorithms on infrastructure-based cameras for a variety of traffic monitoring and situation awareness applications, including connectivity use cases in which the perception capability of CAVs in the surrounding area are supplemented and augmented by the CAROM output.

References

Altekar, N., Como, S., Lu, D., Wishart, J., Bruyere, D., Saleem, F., & Larry Head, K. (2021). Infrastructure-Based Sensor Data Capture Systems for Measurement of Operational Safety Assessment (OSA) Metrics. *SAE Int. J. Adv. & Curr. Prac. in Mobility*, 3(4), 1933-1944. https://doi.org/10.4271/2021-01-0175

Badrinarayanan, V., Kendall, A., & Cipolla, R. (2017). Segnet: A Deep Convolutional Encoder-Decoder Architecture for Image Segmentation. *IEEE Transactions on Pattern Analysis and Machine Intelligence*, 39(12), 2481-2495.

Garlin, C. (2015). (2000) Andrew Ng: Why 'Deep Learning' Is a Mandate for Humans, Not Just Machines. *Wired.* Retrieved from https://www.wired.com/brandlab/2015/05/andrew-ng-deep-learning-mandate-humans-not-just-machines/

Goodfellow, I., Bengio, Y., Courville, A., & Bengio, Y. (2016). *Deep Learning*, Vol. 1, no. 2. MIT Press, Cambridge.

He, K., Gkioxari, G., Dollár, P., & Girshick, R. (2017). Mask R-CNN. *Proceedings of the IEEE International Conference on Computer Vision.* Venice, Italy, 2961-2969.

Kahn, J. (2016). Forward Thinking: March of the Machines. *Bloomberg.* Retrieved from https://www.bloomberg.com/news/articles/2016-05-20/forward-thinking-robots-and-ai-spur-the-fourth-industrial-revolution

Lang, A. H., Vora, S., Caesar, H., Zhou, L., Yang, J., & Beijbom, O. (2019). Pointpillars: Fast Encoders for Object Detection from Point Clouds. *Proceedings of the IEEE/CVF Conference on Computer Vision and Pattern Recognition.* Seattle, WA, 12697-12705.

LeCun, Y., Bottou, L., Bengio, Y., & Haffner, P. (1998). Gradient-Based Learning Applied to Document Recognition. *Proceedings of the IEEE* 86(11), 2278-2324.

Leshno, M., Lin, V. Y., Pinkus, A., & Schocken, S. (1993). Multilayer Feedforward Networks with a Nonpolynomial Activation Function Can Approximate Any Function. *Neural Networks* 6(6), 861-867.

Lu, D., Jammula, V. C., Como, S., Wishart, J., Chen, Y., & Yang, Y. (2021). CAROM—Vehicle Localization and Traffic Scene Reconstruction from Monocular Cameras on Road Infrastructures. *IEEE International Conference on Robotics and Automation (ICRA 2021).* Xi'an, China.

Lucas, B. D., & Kanade, T. (1981, August 28). An Iterative Image Registration Technique with an Application to Stereo Vision. *Proceedings of the 7th International Conference on Artificial Intelligence (IJCAI).* Vancouver, Canada, pp. 674-679.

Ren, S., He, K., Girshick, R. B., & Sun, J. (2015, December). Faster R-CNN: Towards Real-Time Object Detection with Region Proposal Networks. *Proceedings of the 28th International Conference on Neural Information Processing Systems (NIPS).* Montreal, Canada, pp. 91-99.

Smilkov, D., & Carter, S. (n.d.). Tensorflow Playground. Retrieved from https://playground.tensorflow.org/

Szeliski, R. (2010). *Computer Vision: Algorithms and Applications.* Springer Science & Business Media, London.

Thrun, S., & Urmson, C. (2011). The Evolution of Self-Driving Vehicles. *IEEE International Conference on Intelligent Robots and Systems (IROS 2011)*, Keynote Speech. San Francisco, CA.

Tomasi, C., & Kanade, T. (1991). Detection and Tracking of Point. Features. Technical Report CMU-CS-91-132, Mellon University, Carnegie.

Velodyne Lidar. (2019). Velodyne Ultra Puck LIDAR Datasheet. Retrieved from https://velodynelidar.com/wp-content/uploads/2019/12/63-9378_Rev-F_Ultra-Puck_Datasheet_Web.pdf

VOL
MAP
RADIO
MEDIA
NAV
FRONT
REAR
PASSENGER
AIR BAG
A/C
CLIMATE
MODE
DUAL
CleanAir
TEMP
AUTO
TEMP

6

Sensor Fusion

Sensor Fusion Definition and Requirements

Human drivers constantly monitor the vehicle speed and position, the roadway, and the motions of other road users, synthesizing this information to inform the next action or maneuver. Much of the required perception is visual, though there may be auditory (e.g., car horns) or tactile (e.g., highway rumble strips or haptic warnings from active safety systems) components. To achieve similar functionality, CAVs rely on a data stream from a variety of sensors, using *sensor fusion* to generate a comprehensive picture of the vehicle and its environment. This chapter will discuss the various aspects of sensor fusion, both at a more general level and in the context of CAV applications.

The scenario depicted in Figure 6.1 can help contextualize the role that various sensors play in vehicle driving automation and the importance of sensor fusion. As the CAV (red vehicle) approaches the intersection to navigate a left turn, the various sensors employed by the CAV are collecting information about the environment, while HD maps convey details on the intersection geometry and, at the particular moment shown in Figure 6.1, a pedestrian is entering the crosswalk. The pedestrian location could be captured by the vehicle LIDAR while their speed is measured by the RADAR unit. Cameras are utilized to identify the object as a pedestrian. Meanwhile, two other vehicles are approaching the intersection from the north and the east. LIDAR, RADAR, and cameras may each provide information about position, speed, and type of vehicle detected. This information is crucial for safety-critical considerations, such as whether one of the other vehicles may be traveling at a speed, which would indicate they will fail to stop at the sign; whether one of the other

FIGURE 6.1 Driving scenario example.

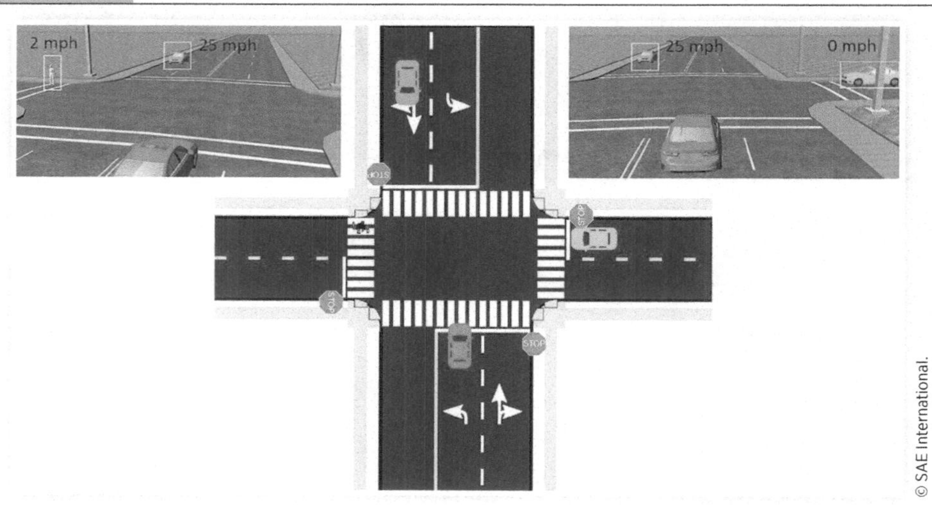

vehicles arrived at the intersection first and therefore has the right of way; or if the observed driving behaviors from the other vehicles warrant extra caution. At the same, time, sensors on board the CAV are used to determine its own speed, position, and orientation. This high-level example highlights the importance of all these sensors working together, combining their data to construct a comprehensive description of the vehicle and its environment for safe CAV operation. In the remainder of this chapter, the fundamentals of CAV sensor fusion models, architectures, and methodologies will be explored in greater depth.

Sensor Fusion Definition and CAV Data Sources

Sensor fusion is the intelligent merging of disparate data from multiple sensors, synthesizing this information in a way that can be used to ensure adequate performance of the system using these sensors. Sensor fusion is currently used in a wide array of applications, from wearables that track steps and exercise to CAVs driving around cities. For any system, different sensors may have advantages and disadvantages for certain applications. For instance, a RADAR unit is effective at directly measuring the speed of a vehicle; however, it is not readily capable of distinguishing a pedestrian from a car. Conversely, a camera is not able to directly measure the speed of a moving object, yet it provides the detailed images necessary to identify and classify objects. Therefore, for complex tasks, like automated driving, that require a combination of parameters to be identified and measured, sensor fusion is essential, utilizing the strengths of certain sensors to supplement and account for the limitations of others to provide the required information necessary for acceptable system performance. Moreover, using multiple sensor modalities with (some) overlapping areas of sensing ability allows for redundancy in cases of sensor error or unusual scenarios.

Chapters 2 through 5 discussed CAV localization, connectivity, sensor and actuation hardware, and computer vision. This chapter will demonstrate how many of the concepts discussed in previous chapters tie together to generate a comprehensive description of the CAV and its environment through sensor fusion. The fusion of data from both on-board and off-board sensors establishes the CAV's world model, defined in SAE J3131 as "*The ADS's internal representation of the portions of the environment of which the system is aware, or that are of interest to the system and/or the user for the purpose of achieving a specific goal*" (SAE International [in progress]). The world model includes both static and dynamic aspects of the surrounding environment. Static features include things that are unlikely to change such as signage and driving laws, while dynamic features, such as other road users, the weather, and road conditions, may constantly change. Both aspects are essential in providing context to the CAV from the mission planning phase through the completion of the DDT. While more data generally lead to improved world models, OEMs may adopt different methods to construct these models based on available sensors, computational and communications hardware, and software development decisions.

In most CAV applications, sensors for estimating the surrounding environment of the vehicle typically include some combination of cameras, LIDAR, and RADAR, though these are just a few of many possible sensors (see Chapter 4) that could contribute relevant information to a *fusion electronic control unit* acting as the "brain" of the CAV, as depicted in Figure 6.2. Employing multiple techniques to obtain and merge data provides redundant functionality, increases the robustness of CAV perception and localization, and improves its overall understanding of the surrounding environment.

FIGURE 6.2 Example configuration of sensors processed by the CAV ECU.

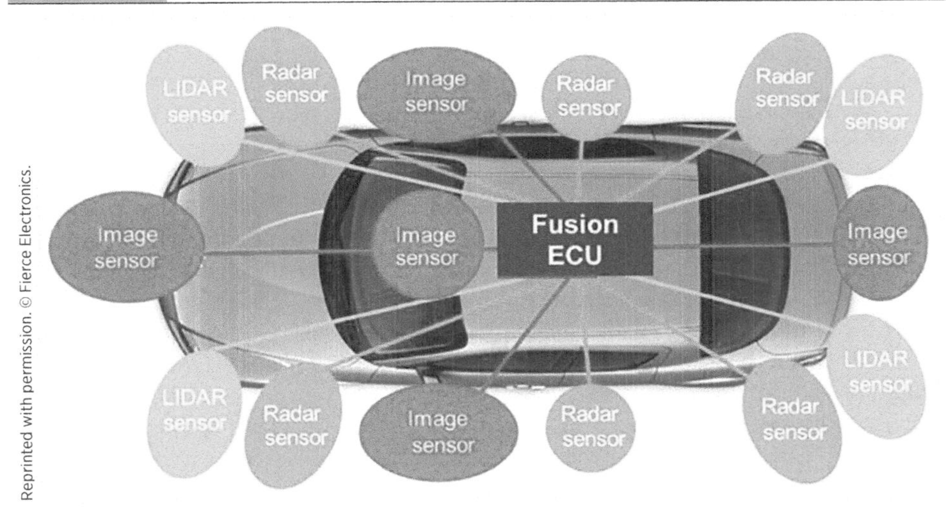

In addition to the real-time sensor data described above, information can be provided to the vehicle in advance for heightened situational awareness. A priori knowledge of the environment can include HD maps (see Chapter 2) with the potential to show information about the immediate surroundings such as lane width, roadway curvature, and the road gradient. By providing these data in advance, the a priori information from HD maps may reduce the real-time computational load on the CAV since the information can be processed and synthesized before the vehicle reaches a given location. It may also obviate the need for certain real-time sensing capabilities, potentially reducing overall hardware costs and/or processing time.

However, there are some disadvantages to relying on such a priori information in highly dynamic environments. For instance, consider how often temporary detours, roadway changes, or closures due to construction or collisions affect travel. Roadways undergo other changes over time as well, such as restructuring of the roadway itself (e.g., restriping or changes to lane geometry) or general degradation (e.g., lane markings becoming difficult to see). Furthermore, there is an aspect of trust involved in relying on provided information regarding the environment that may have been collected by other entities. Since the environment is subject to change and the vehicle cannot validate the data being provided without independent measurements, this information may only be considered supplemental as a redundant source to that which is measured directly by the vehicle. The reliability of a priori information and related cybersecurity issues threatening vehicle connectivity is a major challenge currently faced within the industry for which there has yet to be an agreed-upon solution.

Vehicle state estimation may be improved through the use of past data on vehicle capabilities and trends. Each vehicle on the roadway operates within certain limits depending on its capabilities. For instance, a supercar will have a far different acceleration and braking capacity than a minivan; however, assuming proper maintenance, the capabilities of any given vehicle may be assumed to be relatively consistent and may be estimated to a reasonable degree of certainty depending on the operating condition. As such, the estimation of the vehicle state over time can be refined by updating estimates of key parameters such as vehicle inertial properties and tire friction. If sensor measurements fall outside of the expected range, state and parameter estimation can be updated and improved through well-known techniques such as a Kalman filter.

As always, there is some level of measurement uncertainty (MU) associated with data collected by CAVs. This uncertainty decreases with greater sensory accuracy and precision. Many sensors have a measurement precision that is approximately described by a normal distribution, and sensor specifications may provide a *standard deviation* value. In a normal distribution, there is a 68% chance that the value will lie within one standard deviation of the sample mean, a 95% chance that a given measurement falls within two standard deviations, and a 99% chance that a given measurement is within three standard deviations. The smaller the standard deviation, the greater the sensor precision. On the other hand, sensor accuracy is generally a function of calibration and measurement drift over time. The use of high-precision, properly calibrated sensors is important to ensure that the CAV has access to high-quality information.

In order to better understand the concept of MU as it applies to sensor fusion, consider the example of a GPS sensor, which is a piece of instrumentation that most vehicles have been equipped with for years due to its utility and relative low cost. Anybody who has ever been lost on the way to a destination even with GPS assistance is likely all too familiar with the disadvantages of traditional GPS units. In order to function properly, the GPS requires a satellite connection to triangulate position. However, there are many conditions that may compromise this connection, such as interference from mountain ranges or tall buildings, poor weather conditions, and passing through a tunnel. As a result, the GPS unit may temporarily disconnect and be unable to locate the vehicle. If a CAV were to rely solely on the instantaneous position data from the GPS, it would have no clue as to whether it was in downtown Boston or the middle of the Sahara. Intuitively, a human driver would know that if the last GPS reading a few seconds prior placed the vehicle in downtown Boston, there is a zero possibility that moments later it would be on another continent. This intuition can be mathematically represented in the form of a Kalman filter.

A simple Kalman filter operates under a two-step function of prediction and updated measurement. Using several of the latest data points, the most recent known position of the vehicle can be calculated, along with an estimate of vehicle heading and velocity. Using these results, a prediction can be made for the vehicle's position at the next time step. Once another position reading can be made by the GPS unit, the measurement can be updated and the next state prediction can be calculated. From this example, one can easily see how known information regarding the vehicle performance, environmental conditions such as speed limit, and information derived from other sensors could be used to further refine the state prediction algorithm to lower the uncertainty in vehicle position. Synthesis of data that are sufficiently reliable and robust for CAV operation requires the use of redundant sensors, historical data, vehicle specifications, and a priori knowledge regarding the surrounding environment.

Sensor Fusion Requirements

In order to successfully perform the sensor fusion design objective, it is important to know what information and data are necessary to successfully perform the task at hand. Requirements for the proper application of sensor fusion may include, but are not limited to, the following:

1. The algorithms or techniques must be appropriate for the particular application.
2. The architecture (i.e., where, in the processing flow, the data are fused) must be chosen judiciously.
3. The choice of sensor suite must be appropriate for the particular application.
4. The processing of sensor data must be able to extract the required amount of useful information.
5. The data fusion process requires high precision and accuracy.
6. The data fusion process must have sufficiently low latency for the particular application.
7. There must be an understanding of how the data collection environment (i.e., signal propagation, target characteristics, etc.) affects the processing.

Different applications employing sensor fusion will have different requirements. Some may call for sensors with lower or higher accuracy, precision, and resolution (the constituents of MU) depending on the design objective. While it is important to ensure all available information is appropriately processed to generate the most comprehensive understanding of the vehicle and environment states, data fusion methods should also consider related costs; it is important to create a practical product that offers the desired performance without overdesigning the system.

Sensor Fusion Origins

The development of formal sensor fusion techniques began long before the sustained development of automated vehicles. In fact, the first general data fusion model was developed more than three decades ago in 1985 by the U.S. Joint Directors Laboratories (JDL) data fusion group, now known as the Data and Information Fusion Group for the U.S. Department of Defense. Multi-sensor data fusion techniques were originally developed for aeronautics, surveillance, and defense applications; however, in recent years, sensor fusion has been studied for a broad range of the context of the growing "internet of things" and demonstrates clear applications for automated technologies.

While the JDL model was the first to be introduced focused on a data-based approach, there have since been various techniques adopted. The most widely propagated data fusion can be broadly categorized into data-, activity-, or role-based approaches as follows (Almasri & Elleithy, 2014):

- **Data-Based Model**
  - *JDL Model*
  - *Dasarathy Model a.k.a. Data-Feature-Decision* (*DFD*)
- **Activity-Based Model**
  - *Boyd Control Loop* (*OODA Loop*)
  - *Intelligence Cycle*
  - *Omnibus Model*
- **Role-Based Model**
  - *Object-Oriented Model*
  - *Frankel-Bedworth Architecture*

JDL Model

Although the available sensors and technologies surrounding data fusion have undergone extensive improvements and refinement since the first data fusion model was developed in 1985, the original underlying concepts and data fusion architecture of the JDL model remain unchanged. The primary purpose of the original data fusion model developed by JDL was to establish a unified, common framework for data fusion. This JDL model defined data fusion as (Steinberg & Bowman, 2001):

> A process dealing with the association, correlation, and combination of data and information from single and multiple sources to achieve refined position and identity estimates, and complete and timely assessments of situations and threats, and their significance. The process is characterized by continuous refinements of its estimates and assessments, and the evaluation of the need for additional sources, or modification of the process itself, to achieve improved results.

This definition was established primarily with defense applications in mind. In 2001, Steinberg and Bowman identified modifications to this original definition that broadened the scope for various applications of data fusion (Steinberg & Bowman, 2001). Some of these modifications included:

- De-emphasizing data correlation and association, which is not strictly necessary for state estimation.
- Broadening the scope of state estimation beyond "position and identity estimates".

- De-emphasizing threat assessment and including a more general description of applications that require an assessment of utility or cost associated with a situation.

With these modifications, the relevance of data fusion and the JDL model to applications such as CAVs becomes more apparent, and the JDL model is one of the more popular data fusion models today. It is comprised of sources acting as inputs to the model (Figure 6.3, left), five levels of processing within the data fusion domain and a database management system (Figure 6.3, middle), and a human-machine interface (HMI) allowing for human interaction (Figure 6.3, right).

As the application of the JDL model has moved beyond defense applications over time, the definition of the five levels of data processing has been broadened and generalized. Hence, these definitions can be abstract and esoteric, so examples will be provided for each level with regard to their applicability toward automated vehicles.

Level 0—*Source Preprocessing*, modified by Steinberg and Bowman (2001) as *Signal/Feature Assessment*, uses features that are immediately observable by the utilized sensors to provide initial observations regarding the nearby entities in the environment. In the CAV example, Level 0 refers to the directly observable states such as:

- The instantaneous position of the subject vehicle within its lane of travel.
- The instantaneous velocity of the subject vehicle.
- The present control inputs contributing to the DDT.

FIGURE 6.3 Revised JDL data fusion model (1998 version).

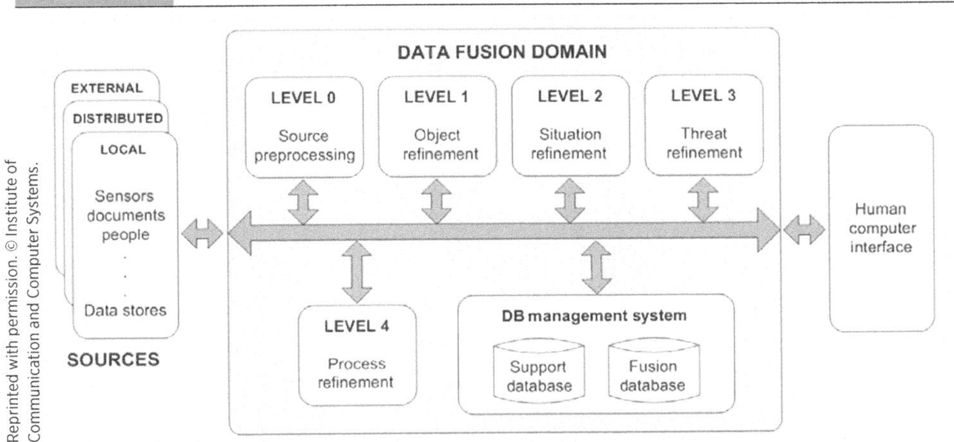

Level 1—*Object Refinement*, revised to *Entity Assessment*, establishes consistent definitions for the various sensor data with the purpose of allowing for estimates of the current and future states. For a CAV, Level 1 involves:

- Identification of surrounding entities in the FOV.
- Classification of surrounding entities helping to identify dynamic capabilities.

- Measured and/or assumed dynamic performances of surrounding entities to predict current and future positions.

Level 2—*Situation Refinement*, updated to *Situation Assessment*, focuses on the relationships between entities that may characterize various situations. Examples of these relationships include:

- The subject vehicle coming to a stop for a red light or stop sign.
- The subject vehicle maintaining the speed limit.
- The subject vehicle slowing down due to a lead vehicle slowing down or stopping.

Level 3—*Threat Assessment*, modified to *Impact Assessment* for a more generalized data fusion approach, focuses on the outcome of a predicted situation. This contrasts with Level 2, which deals with the relationships defining the current situation. In the context of CAVs, Level 3 would be related to safety metrics. Some such metrics could include (see Chapter 8):

- Minimum safe distance—If a minimum safe distance between vehicles is violated, there is a growing likelihood of collision due to this reduced separation distance.
- Time-to-Collision (TTC)—If the trajectories of two vehicles will conflict and cause a collision, TTC is the duration between the present moment and this predicted collision.
- Traffic law violation—A determination that the vehicle is violating a traffic law (e.g., velocity exceeds the speed limit) may result in a heightened risk of conflict.

Level 4—*Process Refinement*, changed to *Performance Assessment*, evaluates the system performance against a defined criterion. This evaluation relates the outputs of the data fusion system back to a ground truth to assess the overall effectiveness of the system. Level 4 in the context of CAVs could include the following types of assessments of vehicle performance:

- Operational safety assessment of the subject vehicle given the results of various safety metrics.
- Sensor accuracy of the subject vehicle when comparing its outputs to a ground truth.
- Behavioral competency assessment of the subject vehicle, comparing the intent of the vehicle system to the measured outcome.

The JDL model forms the basis for many of the other data fusion models in use today. In the following sections, some of the other existing models will be introduced and briefly described, although the core concepts are based on the origins of sensor fusion.

Dasarathy Model

The Dasarathy model is designed to remove ambiguity from data structures by defining levels based on the input and output. The five levels of the Dasarathy model are depicted in Figure 6.4, combining inputs and outputs comprised of raw data, features, and decisions.

FIGURE 6.4 Dasarathy model for data fusion categorization.

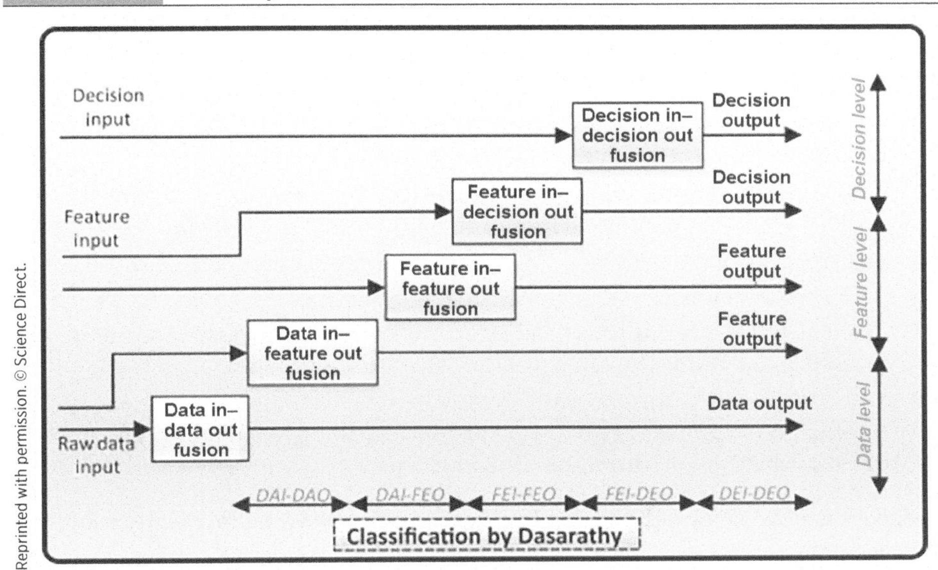

There are different types of fusion operations conducted in the Dasarathy model. *Data in-data -out*-type fusion, for instance, may employ raw data from the RADAR to refine position data collected by the GPS. *Feature in-decision out*-type fusion is performed to transition between groups of levels. For example, a calculation of a minimum safe distance violation (see Chapter 8) at a feature level may result in a decision output to a higher level to slow down the vehicle and avoid a collision.

Boyd Control Loop

As described by Almasri and Elleithy (2014), the Boyd Control Loop, also known as the OODA loop, is a simple, activity-based model for sensor fusion. The loop consists of four stages which can be discussed in the context of CAVs as:

1. **Observe:** The raw information is collected from the available sensors.
2. **Orientate:** Data association allows for the refinement of state prediction.
3. **Decide:** The subject vehicle evaluates the current state in the context of the planned mission for possible conflicts that may require action or continuation of the mission if it is safe to do so.
4. **Act:** The subject vehicle implements the proper driving inputs to carry out the mission or take preventative measures to avoid a conflict.

Although the OODA demonstrates usefulness for certain operations, the architecture is too general for straightforward application to most CAV data fusion tasks, lacking the required technical granularity.

Intelligence Cycle

Another activity-based sensor fusion model explored by Almasri and Elleithy (2014) is the Intelligence Cycle. The five phases are detailed below as applied to CAV sensor fusion:

1. Planning and Direction: This phase would include path planning for the vehicle to determine the route which it would take to complete a trip.
2. Collection: The subject vehicle uses the available sensors to amass information related to the surrounding environment to help navigate the planned path.
3. Collation: The alignment of raw data collected from the available sensors (i.e., data association).
4. Evaluation: During this phase, the data fusion process takes place and metrics can be used to determine the efficacy of the data fusion algorithm.
5. Dissemination: The final phase of the Intelligence Cycle directs the fused information to provide guidance to the vehicle with regard to the necessary driving inputs to safely navigate the planned path.

Like the Boyd Control Loop, the Intelligence Cycle offers a simplified sensor fusion approach that requires significant refinement for most CAV applications.

Omnibus Model

The Omnibus Model presents a greater level of detail than that of the Boyd Control Loop and Intelligence Cycle; however, it employs a similar activity-based strategy (Bedworth & O'Brien, 2000). The OODA steps are taken from the Boyd Control Loop and broken down into substeps, providing additional context to help refine the sensor fusion algorithm. As a result, this model begins to share some similarities with the Dasarathy Model, categorizing the OODA loop into different fusion levels to help reduce ambiguity about expected inputs and outputs for each step.

Object-Oriented Model

Kokar, Bedworth, and Frankel (2000) proposed a role-based, object-oriented model with the goal of establishing a top-down model that can account for system constraints. This model considers the role of each object and their relationships to create a reusable data fusion tool. It can also explicitly account for a human interface within the architecture. This model provides a greater emphasis on the relationship between observations, actions, and the subject system. The importance of these relationships should be considered in the context of CAVs, which often rely on a transparent HMI, dynamic knowledge of the changing environment, and response to given scenarios.

Frankel-Bedworth Architecture

The Frankel-Bedworth architecture is another role-based architecture, relating data fusion concepts to human systems. In this context, the Frankel-Bedworth architecture maps controls to emotions, employing local and global processes to achieve specific goals and milestones through a set of defined processes (Bedworth & O'Brien, 2000).

Sensor Fusion Architecture

To better explain the sensor fusion architecture as it applies to automated vehicles, one may first consider the equivalent processes that take place when a human driver operates a vehicle, depicted in Figure 6.5. A car is equipped with numerous sensors, each of which may be inputs to localized control modules such as the engine control module, the airbag control module, and the powertrain control module. The data from these sensors are also validated and passed up to a vehicle sensor fusion algorithm that collects and synthesizes this information. Some of this information may be conveyed to a human driver in the form of various displays and warnings on the dashboard. The sensory cortex of the human brain can be thought of as the localized data manager of the human system, combining inputs such as vision and hearing from various sensory organs. At the highest level, a data supervisor, which for human-driven vehicles is also essentially the driver's brain, interprets the information and employs a decision-making process to determine the appropriate actions. This process is often represented mathematically using fuzzy logic. In this example, it becomes clear that there are various levels of sensor fusion needed on both a sub-system and overall system level to process all of the available information to achieve suitable performance. In fully automated CAVs, where the contextual experience and awareness of the human is removed from the process entirely, the sensor fusion task becomes far more complicated with many more layers of data requiring fusion processes.

FIGURE 6.5 Sample data fusion architecture applied to a human driving a car.

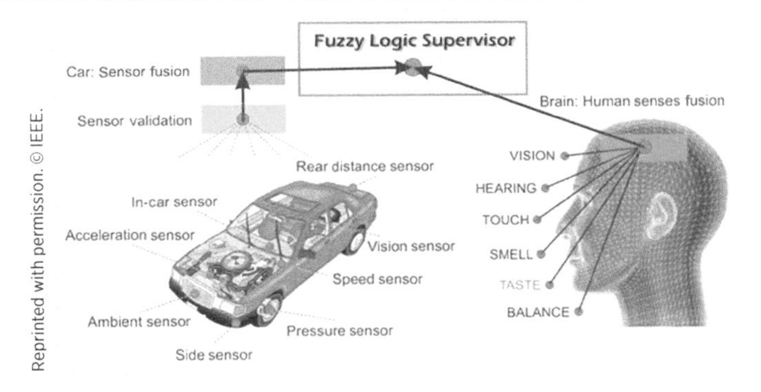

Just as there are several classes of sensor fusion models and algorithms, there are also several different approaches to designing the sensor fusion architecture. There is no single architecture that is fit for all purposes, and the structure of the system should be designed with consideration of the available sensors, the overarching goal of the system, and the limitations constraining the system. The three different types of architectures which will be further examined in the following sections are centralized, distributed, and hybrid fusion architectures.

Centralized Fusion Architecture

Centralized fusion architecture, also known as low-level fusion or signal-level fusion, is fundamentally the simplest of the three architectures; however, the complexity is introduced in the processing and fusion of the raw data (Lytrivis, Thomaidis, & Amditis, 2009). In this type of architecture, all sensor data are processed in one central processor, as illustrated in Figure 6.6; thus, the data fusion algorithm is manipulating the raw data, not processed data. As such, flawed data becomes easy to detect since there are fewer processors acting on data, and all raw data are processed together. It is important to note that the single-point processing of centralized fusion architecture requires spatial and temporal alignment of sensors. If this alignment for any sensor is biased or skewed, raw data from non-corresponding points in space or time could produce erroneous state estimation results.

FIGURE 6.6 Centralized fusion architecture.

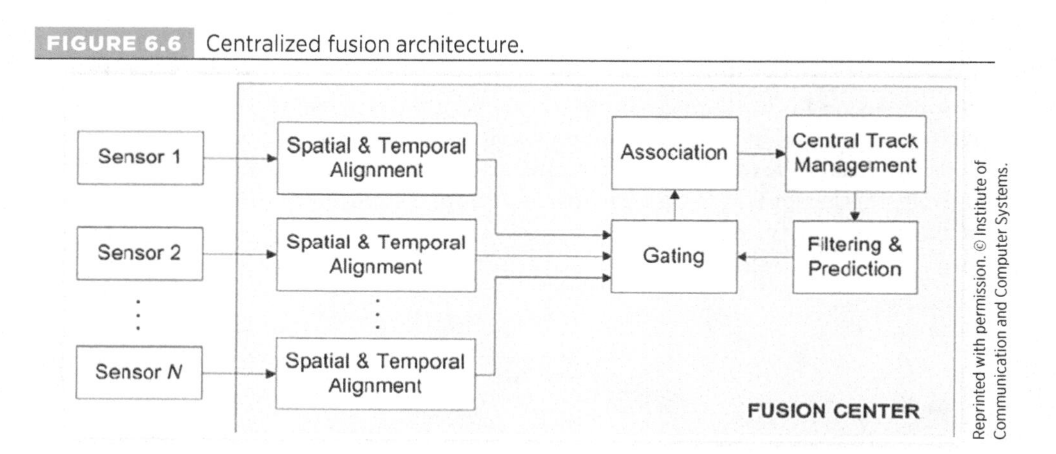

The Multiple Hypothesis Tracking (MHT) algorithm is an exemplar data fusion algorithm often used with a centralized fusion architecture. State estimation using MHT is achieved by either revising the previously predicted state in a process known as track-oriented MHT or by evaluating all possible states tracking global hypotheses by hypothesis-oriented MHT. These techniques are well suited to the centralized architecture because, since raw data are available through a single processor, the fusion algorithm has access to all of the rawdata collected by the system. Although it is useful for the fusion algorithm to have access to all of the raw data, this does create the need for more complex fusion algorithms and computational power since none of the data is preprocessed prior to entering the centralized processor. These and other advantages and disadvantages of centralized fusion architectures are presented in Table 6.1.

Distributed Fusion Architecture

A distributed fusion architecture is also known as high-level fusion or decision-level fusion. In this approach, the input data to the central processor from each sensor are preprocessed.

TABLE 6.1 Centralized fusion architecture advantages and disadvantages.

Centralized Fusion Architecture	
Advantages	**Disadvantages**
Requires the fewest number of processors, so the volume, weight, cost, and power requirements are generally the lowest of the three types.	All sensor data is processed in one central processor, which means that the data fusion algorithm requires the most complexity.
Fewer processors mean fewer processors that can fail, so this can result in increased reliability.	Wide-bandwidth communications (multiple Gbit/s) are required to ensure the central processor receives the sensor data with minimal latency and EMI.
No data are lost due to preprocessing or compression.	The central processor needs to be high powered, which means greater heat generation that must be managed.

The preprocessing activity can occur in the time, frequency, or pixel-based domain. The degree of preprocessing can vary from simple noise filtering to full object identification and classification. Figure 6.7 presents an example of the distributed fusion highlighting the local preprocessing prior to spatial and temporal alignment and fusion.

FIGURE 6.7 Distributed fusion architecture.

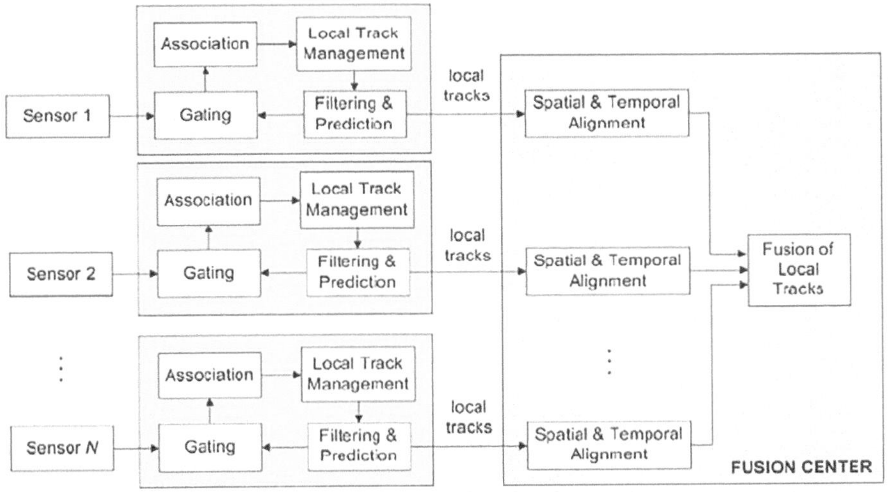

The distributed fusion architecture approach contrasts with the centralized approach, lowering the computational demand on the fusion center by taking advantage of preprocessing steps for portions of the data. Explicit spatial and temporal alignment after preprocessing but before fusion can reduce the likelihood of errors arising from alignment issues. On the other hand, these modules tend to be larger and more expensive than those that employ a centralized approach, and redundancy may be lost if all sensors are not acting independently of one another. Table 6.2 summarizes the advantages and disadvantages of the distributed fusion architecture approach.

TABLE 6.2 Distributed fusion architecture advantages and disadvantages.

Distributed Fusion Architecture	
Advantages	**Disadvantages**
Preprocessing the data puts less demand on the central processor, less power, and heat generation.	Sensor modules become bigger, more expensive, and require more power.
The particular characteristics of different sensor types are used more efficiently.	Functional safety requirements for the sensors are now higher.
Optimization of the preprocessing can occur for each sensor type.	Potential loss of redundancy of the data. Assumption is made that all sensors are acting independently, which may not be the case.
Least vulnerable to sensor failure of the three types.	
Least sensitivity to sensor data misalignment.	
More flexibility in sensor number and type. Changes can be made to the sensor array—addition, subtraction, change of type—without large changes to the data fusion algorithm within the central processor.	
Lower-bandwidth communications are necessary (< Mbit/s).	

Hybrid Fusion Architecture

Hybrid fusion architecture, or feature-level fusion, combines features such as edges, corners, lines, textures, and positions into a feature map that can then be used for segmentation and detection. This approach adopts some strengths of both the centralized and distributed architectures, increasing efficiency while maintaining redundancy. As shown in Figure 6.8, these advantages come at the cost of greater architectural complexity. The hybrid architecture also has the potential for high data transfer requirements due to cross-correlation of local and central tracks.

FIGURE 6.8 Hybrid fusion architecture.

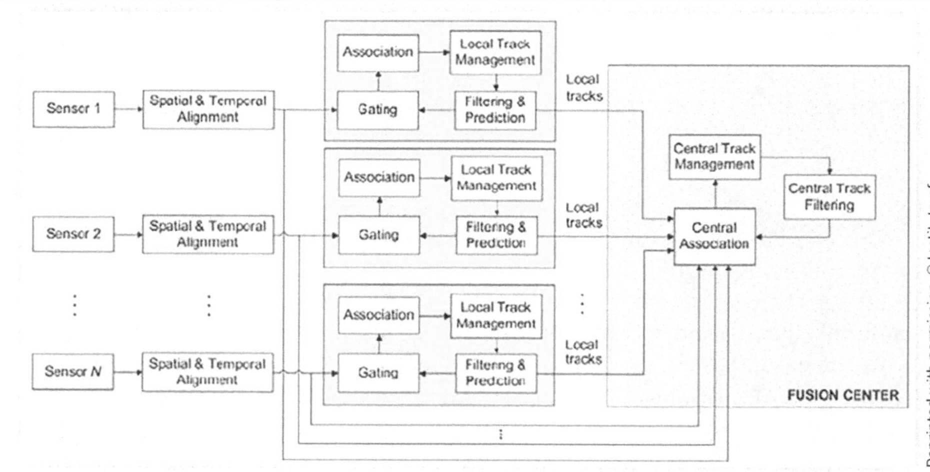

Sensor Interaction

The required sensor modalities and interactions depend on the data fusion objective, which, in turn, relies on the overall mission objective of the system. The primary sensor modalities utilized in CAVs were described in greater detail in Chapter 4. The outputs for these sensors are summarized in Table 6.3. Many of the sensor modalities provide a means to directly measure position, while RADAR units allow the only direct and accurate measure of velocity. There are no direct measures of acceleration depicted (accelerometers may measure the acceleration of the host vehicle, but not of any other road users). When estimating secondary states from directly observable ones, there is a risk of error propagation. For example, using GPS to measure position will inherit some error; however, using GPS to measure acceleration will result in a much greater error due to limitations in resolution and the combination of errors propagated through the different operations. In such complex systems, it is important to incorporate redundant measurements, utilize proper sensor modalities to most directly measure the desired output, and understand the overall MU of the system.

TABLE 6.3 Sensor modality outputs.

Sensor	Measurement	Output description
Video Camera	Position	Image data composed of pixels identifying various traffic actors measured at a frequency consistent with the camera frame rate.
RADAR	Velocity	Speed of traffic actors.
LIDAR	Position	Point cloud data defining the boundary of nearby traffic actors as well as the environment.
GPS	Position	Position for the subject vehicle within the global coordinate system.

Interactions between sensors can help reduce overall MU in a few ways. First, utilizing multiple sensor modalities can improve sensor effectiveness across conditions (e.g., cameras do not provide much information without proper lighting, whereas LIDAR units can provide point cloud data regardless of lighting conditions). Next, the use of multiple sensors may be employed to validate the measured data in case of failed sensor hardware or software. Lastly, different sensors may be better suited to specific aspects of a given data fusion objective. For instance, LIDAR may successfully identify the location of objects, but images from a camera may be necessary to classify the objects. These various interactions between sensor modalities are more formally broken into three categories (Galar & Kumar, 2017), as illustrated in Figure 6.9.

- **Complementary:** Sensors perceive complementary information by covering different areas. Advantages include expanded spatial and temporal coverage, resulting in improved detection of objects.
- **Competitive:** Sensors perceive the same information by covering the same area. Advantages include increased dimensionality, object detection reliability, and spatial resolution, as well as redundancy.
- **Collaborative:** Involves a combination of complementary and competitive approaches in which the sensors partially perceive the same information by covering part of the same area. The advantage of this approach primarily includes more information is perceived than through a single sensor.

FIGURE 6.9 Sensor interaction structure.

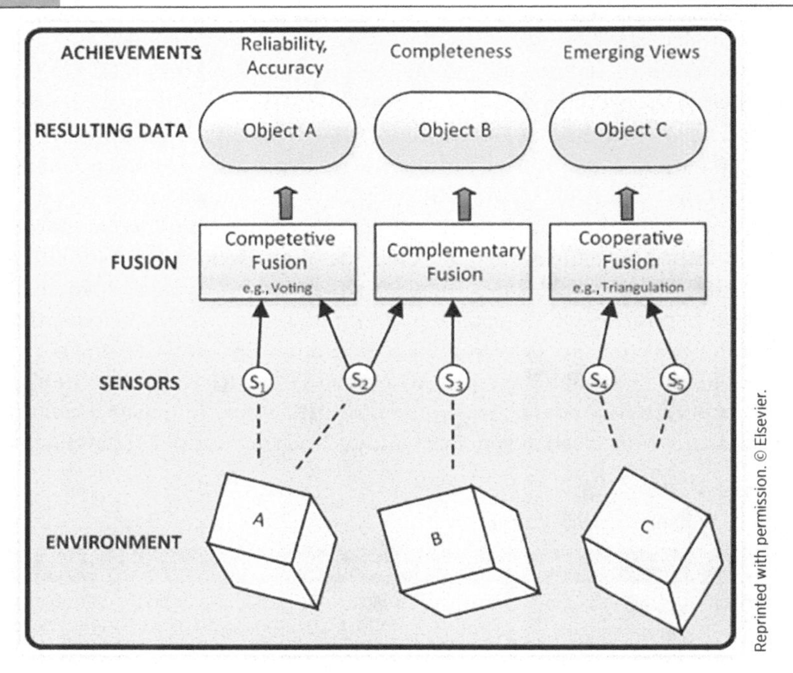

Figure 6.10 offers a visual example of sensor interaction, presenting views from cameras and LIDAR located within an intersection infrastructure. The lower image shows a point cloud of the environment surrounding the vehicle, identifying how close objects in the surrounding environment are to the vehicle without including photographic level details. The upper right image depicts a camera image of the surrounding area providing detailed features of the objects and environment. Images such as this are commonly used to train AI algorithms and can therefore be used to classify objects as vehicles, pedestrians, signage, etc. The upper left image provides a zoomed view that can be better utilized to identify detailed features and help classify the pedestrian crossing the street, facilitating the LIDAR data which provides accurate positional data; however, it may not provide the resolution needed to perceive the entity as a pedestrian.

Alone, any one of these views may not provide a complete picture of the surrounding environment and is insufficient to determine necessary actions to safely operate a CAV; however, combining these sensor modalities through a compilation of complementary, competitive, and collaborative sensor interactions creates a comprehensive overview of the vehicle surroundings to guide the decision-making processes.

FIGURE 6.10 Sensor views from an automated vehicle.

Object and Situation Refinement Examples

Feature Extraction

Recent advances in deep learning and AI have demonstrated great promise in facilitating the successful operation of CAVs. Sensor fusion plays a major role in these processes and is essential to the ability of a vehicle to detect objects in the surrounding environment, understand the actions of such objects, and provide context as to the intent of these objects at future points in time. One way in which sensor fusion assists in deep learning is through feature extraction. Objects can be decomposed into simplified depictions using common geometric shapes. With some basic prior knowledge about common object shapes, feature extraction can separate distinct objects. This prior knowledge can be generated through supervised learning or other methods described in Chapter 5. Fusion of LIDAR and camera data can enable precise object location in 3D space and accurate object identification. For example, this process could include determining a road surface with a point cloud captured by LIDAR data, removing the points belonging to this road surface, and identifying the remaining objects on the road surface as objects in the environment, as depicted in the series of images in Figure 6.11.

FIGURE 6.11 Feature extraction utilizing LIDAR data.

(a) Raw laser point cloud projected on an Image

(b) Ground plane detection on the horizontal plane in point cloud data

(c) Ground plane removal and projection of laser points on vertical surfaces

Multi-Target Tracking

Multi-Target Tracking (MTT) employs multiple detections over time with constant updates of the locations of detected objects in space and predictions of where the objects will be in future frames (Yenkanchi, 2016). Once an object is identified in space, estimations of future states can be established through measurements associated with the current state. The predicted state for each time step is known as data association. Every measurement is assigned to exactly one predicted point trajectory, and every predicted state is assigned to one measurement at most. In the context of CAVs, MTT is best suited for tracking other traffic actors around the subject vehicle. Multiple sensors may be employed for such tracking. LIDAR may be used to determine the 3D location of nearby pedestrians and vehicles, camera-based images may be used to classify the objects into a vehicle or pedestrian, and RADAR may be utilized to determine the real-time speed of the objects. Again the fusion of these data becomes critical in order to locate the current position, determine the capabilities and characteristics of the objects, and estimate future states using the present conditions. Figure 6.12 illustrates the method of MTT by associating measurements to predicted states of the surrounding objects.

FIGURE 6.12 MTT diagram.

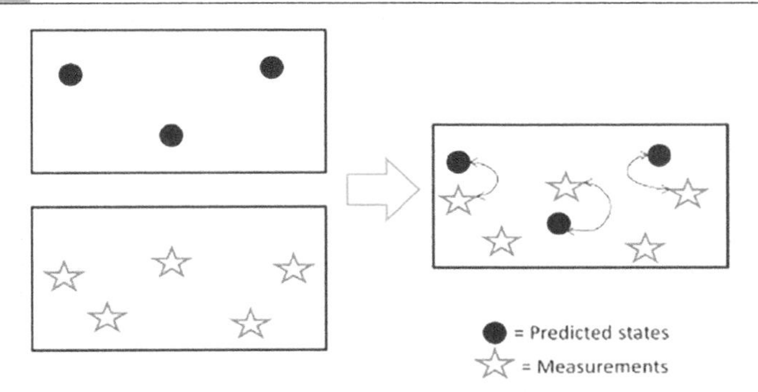

Reprinted with permission. © Shashibushan Yenkanchi.

Evaluation Metrics

Sensor fusion is only a useful technique if the results can be quantified and validated. For CAVs, if a vehicle classifies detected objects with high accuracy but only detects half of the surrounding objects to begin with, the system does not effectively provide the information to safely navigate the surrounding environment. Various evaluation metrics can be applied to sensor fusion measurements to quantify the efficacy of the overall system. Such evaluation metrics include but are not limited to:

- **Sense:** Detection rate, the ratio of true positives to the sum of true positives and false negatives.

$$\text{Sense} = \frac{\text{True Positives}}{\text{True Positives} + \text{False Negatives}}$$

- **Precision:** Positive prediction, the ratio of true positives to the sum of true positives and false positives (i.e., all detected objects).

$$\text{Precision} = \frac{\text{True Positives}}{\text{True Positives} + \text{False Positives}}$$

- **Weighted harmonic mean**: Also known as the F-measure or figure of merit:

$$F - \text{measure} = \frac{2 \cdot \text{Sense} \cdot \text{Precision}}{\text{Sense} + \text{Precision}}$$

- **False positive rate:** Ratio of false positives to the total number of objects detected.

$$\text{False Positive Rate} = \frac{\text{False Positives}}{\text{Total Number of Detected Objects}}$$

- **False negative rate:** Ratio of false negatives to the total number of objects detected.

$$\text{False Negative Rate} = \frac{\text{False Negatives}}{\text{Total Number of Detected Objects}}$$

- **Intersection over Union (IoU):** Ratio of the area overlap between the detection by two sensor modalities of an object to the area union between the detection by two sensor modalities of an object.

$$\text{IoU} = \frac{\text{Area of Overlap}}{\text{Area of Union}}$$

Figure 6.13 illustrates a tracking algorithm which provides speed data for vehicles in the frame; however, the car near the center of the frame has not been tracked. This is an example of a false negative. False negatives can be caused by factors such as occlusions, faulty sensors, or an insufficiently trained data set. A successful sensor fusion algorithm for CAVs will have acceptable values for all these evaluation metrics described, but what is deemed acceptable depends on the specific function. Nevertheless, any algorithm should be evaluated for such metrics to better understand their efficacy prior to being employed in on-road automated vehicles.

FIGURE 6.13 Illustration of false negatives in detection algorithm.

Sensor Fusion Applications: Active Safety Systems

There are features on vehicles today that employ the types of sensor fusion approaches that will be common in CAVs, though they are limited in scope. Many are used for active safety systems or other driving aids, which are often seen as stepping stones to fully automated vehicles. Some active safety systems features such as AEB are likely to see rapid widespread adoption; they were mandatory in new cars from 2020 onward in Europe, and manufacturers have agreed to voluntarily include this feature in the U.S. market starting in 2022. Figure 6.14 summarizes the active safety systems features and driving aids available on many of today's vehicles.

FIGURE 6.14 Sensor diagram for modern CAVs.

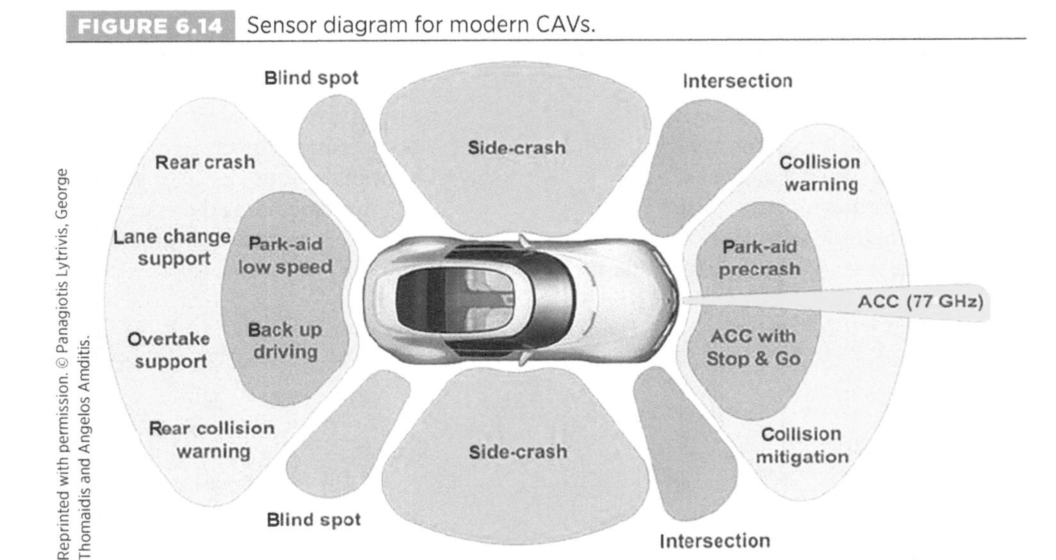

Safe Speed and Distance

Excessive speed has been identified as the cause of approximately 30% of fatal crashes in Europe (European Commission, 2021) and 26% of fatal crashes in the United States (NHTSA, 2018). Longitudinal support systems such as ACC can assist in maintaining a safe speed and distance between vehicles and react to other objects on the road. These systems utilize both long-range and medium-range RADAR to detect objects, while cameras may be employed to detect lane markings as detailed in Figure 6.15. Simultaneously, differential GPS and inertial sensor data are fused with map data to enhance positioning determination. In the context of determining safe speed and distance for a CAV, the prediction of both the subject vehicle path and those of surrounding vehicles is vital.

FIGURE 6.15 Sensor fusion application in ACC.

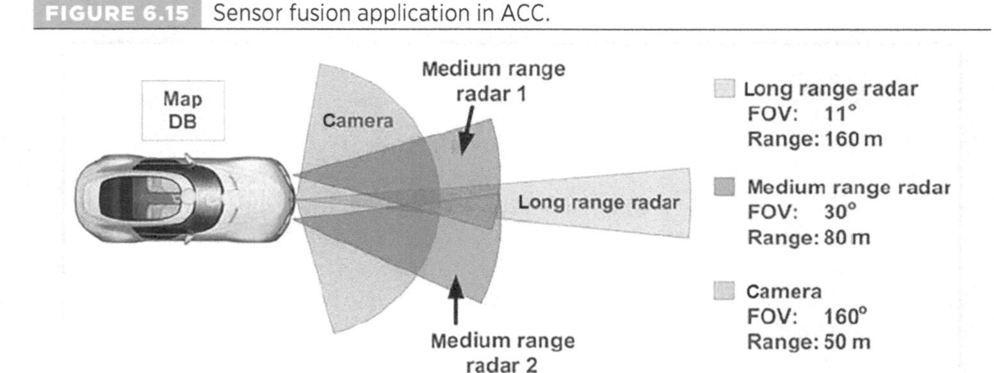

Lane-Keeping Assist

Lateral support systems assist in maintaining the vehicle within its lane of travel. Such systems include an actuator that assists with steering and/or executes a warning that may be haptic, audible, and/or visual to alert the driver of a lane departure. Low latency processing is crucial for LKA systems as lane departures can occur suddenly and pose an immediate threat to both the subject vehicle and nearby road users. In many modern LKA systems, RADAR identifies moving obstacles and cameras monitor the lane, including road curvature. The latter are fused with map data as a redundancy. Figure 6.16 demonstrates the application of sensor fusion for use in LKA systems.

FIGURE 6.16 Sensor fusion application in LKA systems.

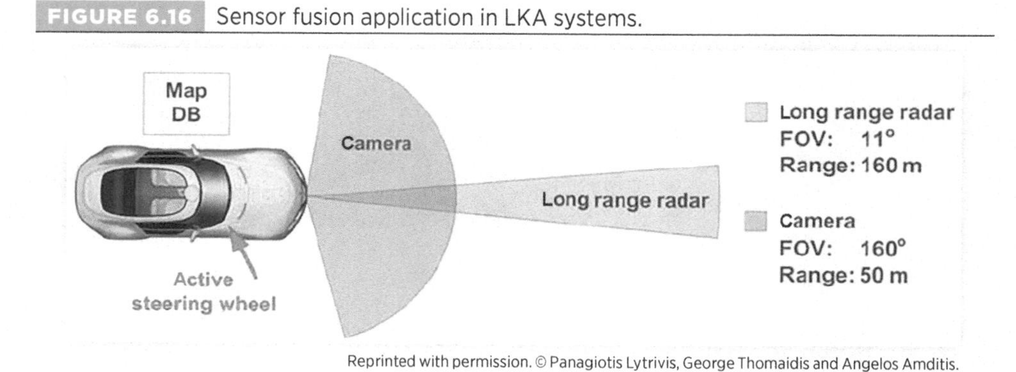

Intersection Navigation

In Europe, approximately 40% of all accidents with major injuries and 18% of all fatalities occur at intersections (Wisch, et al., 2019). These numbers are similar to those reported in the United States (FHWA, 2020). Intersection safety systems are often designed to protect vehicles and more vulnerable entities such as pedestrians and cyclists. Some vehicles on the road today have AEB systems that are designed to react to potential conflicts while in an intersection. However, these systems are not yet designed to autonomously navigate an intersection. To enable this functionality, systems may employ LIDAR for object detection and positioning, cameras for detection of lane markings and traffic signals, as well as object classification, and GPS for the localization of vehicles and comparison against a detailed mapping of the intersection. Figure 6.17 demonstrates the use of LIDAR and camera data providing complementary sensors for successful intersection navigation.

FIGURE 6.17 Sensor fusion application for intersection navigation.

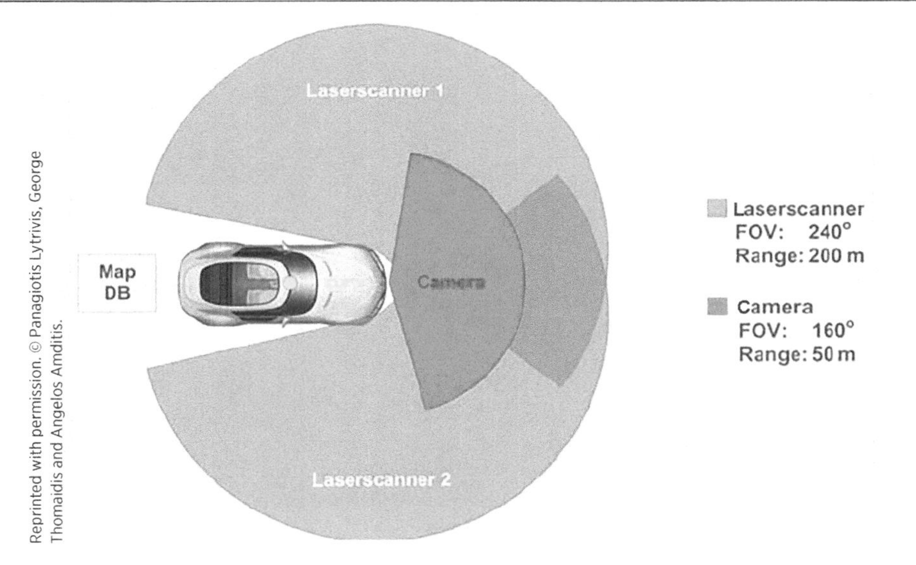

In addition to on-board vehicle sensors, vehicle connectivity provides another data source with the potential to significantly increase the awareness of a CAV in an intersection environment. Connectivity can inform passing vehicles through the using of SPAT data, other vehicle intended actions through an intersection manager, and even help direct CAVs safely through an intersection based on the actions of the surrounding traffic participants. Although connectivity poses numerous technical and legal challenges to the industry and is, therefore, a somewhat controversial topic, there are clear advantages to having access to data collected by other road users and infrastructure-based sensors, even if just for the sake of redundancy and robustness.

Sensor Fusion Examples from Developmental CAVs

The CAV industry is rapidly evolving as vehicle manufacturers continue to optimize the technology needed to safely navigate public roadways. While improvements to sensor accuracy and precision, sensor cost reductions, and supporting AI development for sensor fusion algorithms are all helping drive the industry forward, CAV developers are adopting different sensor suites and strategies to achieve their goals.

Waymo Sensor Fusion Applications

Waymo has long been recognized as one of the leaders in CAV development, and they attribute much of that credit to their in-house approach in hardware and software design. Like many other CAV manufacturers, Waymo employs a combination of LIDAR, RADAR, and camera data to identify objects in the surrounding environment and determine the actions needed to safely operate, as shown in Figure 6.18. The fifth-generation Waymo "driver" utilizes five LIDAR units, four of which are short-range perimeter LIDAR sensors

FIGURE 6.18 Fifth-generation Waymo sensor diagram.

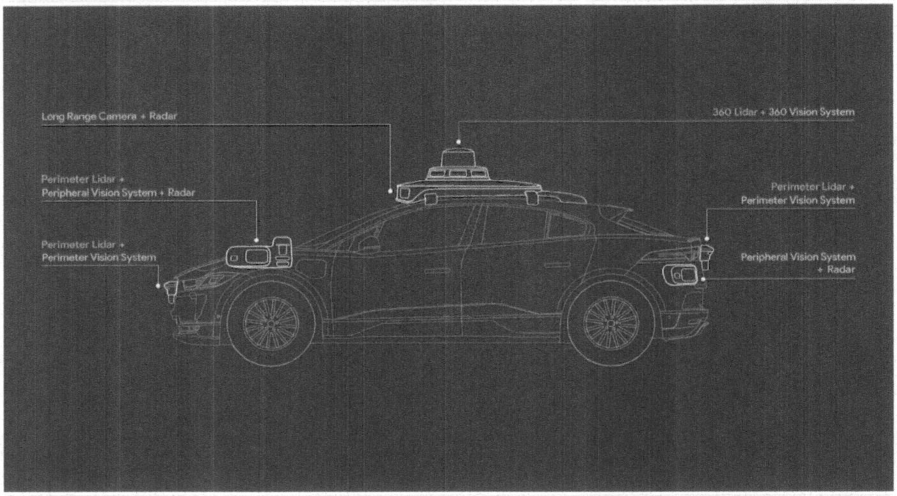

providing close-range detection for objects within the surrounding environment. The 360° LIDAR unit on the top of the vehicle provides long-range data for objects up to several hundred meters away. Waymo's long-range cameras provide photographic vision greater than 500 m away.

Lyft Self-Driving Platform with Level 5

Lyft's CAV platform employs LIDAR, RADAR, and cameras through complementary sensor interactions. While the sensor suite is comparable to that employed by other companies, Lyft has pursued a different sensor fusion approach. It is exploring combinations of early and late fusion techniques to provide a set of baseline data. These data are then carried through to the final perception output, and the results are used to tune the future inputs using early fusion techniques that are capable of fusing the raw sensor data. It should be noted that in 2021, Lyft sold its CAV division to Toyota.

Cruise Application of Late Fusion Techniques

Like the other CAV developers out there, Cruise is exploring its own sensor fusion techniques to develop its CAVs. Cruise is using a late fusion approach in which the raw data captured by each sensor are processed independently for tracking and then fused afterward in the data association phase to improve the result. This process is achieved, for example, by the independent processing of LIDAR and camera data to achieve 3D bounding boxes before being fused at a later stage.

These few examples themselves illustrate the complexity and diversity of sensor fusion techniques currently under development for CAVs. While there are well-established methods for applying data fusion, there is no single answer as how to best inform a vehicle of its surroundings and in real time, nonetheless. There are commercial incentives for companies to innovate and develop new data fusion methods in the race to design a first-to-market CAV, but as the industry moves forward, it is likely that CAV developers will converge on a proven methodology as it is discovered, similar to the progression of the traditional automobile throughout history.

Sensor Fusion Challenges

The application of sensor fusion to CAVs is not without significant challenges. One of the greatest obstacles is the balance between collecting and processing all the required information while doing so efficiently enough for real-time processing. Information overload can lead to insurmountable data processing requirements that are burdensome for the real-time calculations necessary for automating the driving task. Another potentially less obvious challenge is the fact that both stationary and moving objects can create conflicts for a CAV. Not only do sensors need to accurately perceive moving objects but they need to just as accurately sense stationary objects along the road, detect road markings, and interpret signage.

Roadway infrastructure is not always well maintained and may have degraded traffic control devices. Faded stop signs and lane markings, broken traffic signals, or newly paved roads that have not yet been repainted are all situations that experienced drivers are capable of maneuvering but may be problematic for a CAV that is not trained for such situations. The following examples of incidents may highlight considerations for CAVs with regard to sensor information and fusion processes. CAV developers should consider how transparency around the evaluation metrics such as those discussed earlier in the chapter can build public confidence in the capabilities of automated technologies.

While the industry has made great strides toward improving technology and increasing the emphasis on safety as it relates to CAVs on public roads, sensor fusion approaches require continuous development to further optimize CAV operation in difficult environments. Several of the areas in sensor fusion and classification algorithms needing perhaps the most attention include:

- **Occluded Objects:** Objects that become either partially or completely obscured by another object or portion of the environment pose challenges to existing tracking algorithms; however, this is a common challenge in dynamic traffic environments that must be solved for adequate tracking in complex situations.
- **Dynamic Environments:** Many of the current test sites for CAVs include states with consistently clear weather such as Arizona and California. Obviously, CAVs cannot be constrained forever to regions with nice weather, and as such, it is important to develop sensor fusion algorithms which utilize the strengths of a variety of sensors capable of performing in all weather and lighting conditions.
- **Edge Cases:** These scenarios include complex interactions at the "edge" of the vehicle's capabilities. It is important to evaluate sensors and sensor fusion algorithms in such complicated situations to understand system limitations to either restrict the ODD of the vehicle or continuously improve the system to be capable of handling such edge cases.
- **Sensor Failure:** Sensor fusion algorithms are only as good as the sensors used to collect the relevant information. In the case of failures, it is first important to identify whether the issue lies within the hardware or the software, then understand the cause of the failure, and develop solutions.

Lessons from Active Safety Systems and CAVs

Experience with already-deployed active safety systems offers useful lessons for sensor fusion development in CAVs. In 2018, the Insurance Institute for Highway Safety (IIHS) conducted a test series for evaluating active safety systems features on various vehicles including a 2017 BMW 5 Series, 2017 Mercedes-Benz E Class, 2018 Tesla Model 3, 2016 Tesla Model S, and 2018 Volvo S90. This testing included an evaluation of ACC, AEB, and LKA under different scenarios. Such tests are essential to providing the previously described evaluation metrics necessary to consider the efficacy of these systems prior to implementation on public roadways. This testing also offers insight into the sensor fusion capabilities of the system by measuring false positives and false negatives under varying conditions such as the LKA testing which was evaluated on curves and hills with the results summarized in Figure 6.19.

FIGURE 6.19 IIHS test series evaluating LKA on hills and curves.

How active lane-keeping systems performed in IIHS road tests on 3 curves and 3 hills

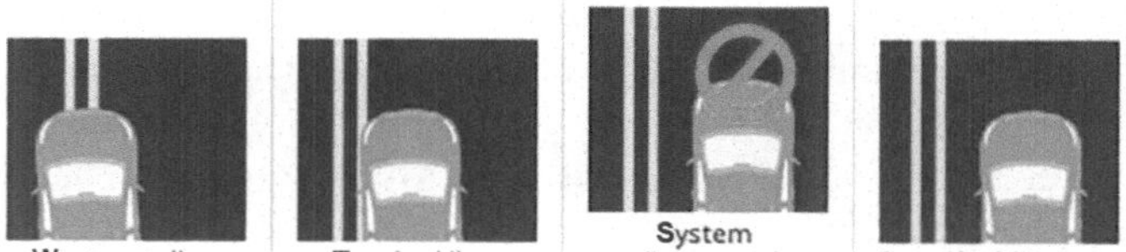

Number of times vehicle

	Went over line		Touched line		System disengaged		Stayed within lane	
	On curves	On hills	On curves	On hills	On curves	On hills	On curves	On hills
BMW 5 series	3	6	1	1	9	7	3	0
Mercedes-Benz E-Class	2	1	5	1	1	1	9	15
Tesla Model 3	0	0	0	1	0	0	18	17
Tesla Model S	1	12	0	1	0	0	17	5
Volvo S90	8	2	0	1	0	4	9	9

Crashes involving CAVs have received significant media attention in recent years. In 2018, a Tesla vehicle collided with a previously damaged crash attenuator on US-101 in Mountain View, California. According to the National Transportation Safety Board (NTSB), the Tesla entered the gore point of an off-ramp at highway speeds, striking the crash attenuator (NTSB, 2018a). The vehicle data reported activation of the Autopilot system at the time of the crash, which is considered a Level 2 automation feature based on the taxonomy of SAE J3016. It is unclear why the driver did not react, and it is likely that the vehicle was unaware of the impacted crash attenuator as the system did not provide pre-crash braking or evasive steering. Around the same time as the Tesla incident, an automated XC90 SUV operated by Uber as a test vehicle struck a pedestrian at nighttime. According to the NTSB final report for this crash, the ADS did properly detect the pedestrian well before impact; however, the vehicle did not properly classify the pedestrian and did not predict her path (NTSB, 2018b). The specifics as to whether or not this resulted from an error in the sensor fusion algorithm are unclear, though it does highlight the importance of the sensor fusion processes to not only identify but also accurately perceive the surrounding objects and ensure the safety of all roadway users. It is important to note that in the cases described, driver intervention should have taken place to avoid the impending collisions as these vehicles relied on driver fallback which was not employed, yet these instances highlight important considerations for sensor fusion applications.

Summary

This chapter presented sensor fusion in both a general sense and as it relates specifically to CAVs. By now, the importance of sensor fusion in developing a safe CAV capable of fully understanding the surrounding environment should be clear. Sensor fusion synthesizes the vast amounts of information required to successfully plan and execute trajectories, safely maneuvering the CAV in complex environments. Although there are many approaches to developing a successful sensor fusion system for a CAV, not all of which have been discussed here, it is crucial that any approach that is taken is comprehensively evaluated to assess limitations and adjudicate conflicts prior to deployment.

References

Almasri, M.M., & Elleithy, K.M. (2014). Data Fusion Models in WSNs: Comparison and Analysis. *Proceedings of the 2014 Zone 1 Conference of the American Society for Engineering Education*. Bridgeport.

Altekar, N., Como, S., Lu, D. et al., Infrastructure-Based Sensor Data Capture Systems for Measurement of Operational Safety Assessment (OSA) Metrics. SAE Technical Paper 2021-01-0175, 2021, doi:10.4271/2021-01-0175.

Bedworth, M., & O'Brien, J. (2000). The Omnibus Model: A New Model of Data Fusion?. *IEEE Aerospace and Electronic Systems Magazine*, Volume 15, Issue 4, pp. 30-36.

Cohen, J. (2020, November 26). Cruise Automation—A Self-Driving Car Startup. Retrieved from https://medium.com/think-autonomous/cruise-automation-a-self-driving-car-company-4dee84ace02d

European Commission. (2021, September 5). Mobility and Transport: Road Safety. Retrieved from https://ec.europa.eu/transport/road_safety/specialist/knowledge/young/magnitude_and_nature_of_the_problem/characteristics_of_these_crashes_en

FHWA. (2020, October 27). Intersection Safety. Retrieved from https://highways.dot.gov/research/research-programs/safety/intersection-safety

Frankel, C.B., & Bedworth, M.D. (2000). Control, Estimation, and Abstraction in Fusion Architectures: Lessons from Human Information Processing. *Proceedings of the Third International Conference on Information Fusion*. Paris.

Galar, D., & Kumar, U. (2017). Chapter 4: Data and Information Fusion From Disparate Asset Management Sources. Edited by Galar and Kumar. In *eMaintenance: Essential Electronic Tools for Efficiency*. Elsevier, London, pp. 179-234.

IIHS. (2018, August 7). IIHS Examines Driver Assistance Features in Road, Track Tests. Retrieved from https://www.iihs.org/news/detail/iihs-examines-driver-assistance-features-in-road-track-tests

Jeyachandran, S. (2020, March 4). Introducing the 5th-Generation Waymo Driver: Informed by Experience, Designed for Scale, Engineered to Tackle More Environments. Retrieved from https://blog.waymo.com/2020/03/introducing-5th-generation-waymo-driver.html

Kokar, M.M., Bedworth, M.D., & Frankel, C.B. (2000). Reference Model for Data Fusion Systems. *Proceedings Volume 4051, Sensor Fusion: Architectures, Algorithms, and Applications IV*. Orlando.

Lytrivis, P., Thomaidis, G., & Amditis, A. (2009). Chapter 7: Sensor Data Fusion in Automotive Applications. Edited by Nada Milisavljevic. In *Sensor and Data Fusion*, I-Tech, Vienna, pp. 123-140.

NHTSA. (2018). Risky Driving—Speeding. Retrieved from https://www.nhtsa.gov/risky-driving/speeding#:~:text=For%20more%20than%20two%20decades,26%25%20of%20all%20traffic%20fatalities

NHTSA. (2019). Federal Motor Vehicle Safety Standard No. 111, Rear Visibility.

NTSB. (2018a). Collision between a Sport Utility Vehicle Operating with Partial Driving Automation and a Crash Attenuator. NTSB, Mountain View.

NTSB. (2018b). Highway Accident Report: Collision Between Vehicle Controlled by Developmental Automated Driving System and Pedestrian. NTSB, Washington, DC.

Rezaei, M., & Fasih, A. (2007). A Hybrid Method in Driver and Multisensor Data Fusion, Using a Fuzzy Logic Supervisor for Vehicle Intelligence. *International Conference on Sensor Technologies and Applications.* Valencia.

Rudoph, G., & Voelzke, U. (2017, November 10). Three Sensor Types Drive Autonomous Vehicles. Retrieved from https://www.fierceelectronics.com/components/three-sensor-types-drive-autonomous-vehicles

SAE International. (In progress). J3131—Definitions for Terms Related to Automated Driving Systems Reference Architecture. SAE International.

Siemens. (2021). Deep Sensor Fusion for Perception and Navigation of Complex Real-World Driving Scenes. Retrieved from https://www.plm.automation.siemens.com/global/pt/webinar/scene-understanding-for-autonomous-vehicles/77102

Steinberg, A.N., & Bowman, C.L. (2001). Chapter 2: Revisions to the JDL Data Fusion Model. Edited by David Hall and James Llinas. In *Handbook of Multisensor Data Fusion*, CRC Press, London.

Vincent, E., & Parvate, K. (2019, December 5). Leveraging Early Sensor Fusion for Safer Autonomous Vehicles. Retrieved from https://medium.com/lyftself-driving/leveraging-early-sensor-fusion-for-safer-autonomous-vehicles-36c9f58ddd75

Wisch, M., Hellmann, A., Lerner, M., Hierlinger, T., Labenski, V., Wagner, M., … Groult, X. (2019). Car-to-Car Accidents at Intersections in Europe and Identificationo of Use Cases for the Test and Assessment of Respective Active Vehicle Safety Systems. *26th International Technical Conference on the Enhanced Safety of Vehicles (ESV): Technology: Enabling a Safer Tomorrow.* Eindhoven.

Wishart, J., Como, S., Elli, M., Russo, B. et al., "Driving Safety Performance Assessment Metrics for ADS-Equipped Vehicles," SAE Technical Paper 2020-01-1206, 2020, doi:10.4271/2020-01-1206.

Yenkanchi, S. (2016). Multi Sensor Data Fusion for Autonomous Vehicles. University of Windsor.

VOL
MAP
RADIO
MEDIA
NAV
FRONT
REAR
PASSENGER AIR BAG
A/C
CLIMATE
MODE
DUAL
CleanAir
TEMP
AUTO
TEMP

7

Path Planning and Motion Control

Definition and Hierarchy

In the context of CAVs, path planning refers to the task of searching for and defining a path or trajectory from the vehicle's current position and orientation to some future goal position and orientation. Typically, *path planning* is considered distinct from the more strategic *route planning*, which is the task of finding an optimal or near-optimal route between an origin and destination address considering factors such as roadway construction, traffic, and weather conditions. Researchers and engineers have continued to develop and refine route planning methods that offer improved efficiency and robustness, leading to algorithms that can find a route or reroute within milliseconds and convey turn-by-turn navigation instructions to human drivers (Bast, et al., 2016). Some of the familiar examples of the latter are easily accessible smartphone apps like Google Maps, Apple Maps, and Waze.

Path planning addresses the challenge of navigating the desired route in a complex, dynamic, and changing environment. In CAVs, the entire decision and control process may take place without human intervention. Many approaches to construct this process involve breaking it down into more manageable subtasks (Paden, Čáp, Yong, Yershov, & Frazzoli, 2016), such as the hierarchical formulation shown in Figure 7.1. The route planning step takes as inputs the user's desired destination and current mapping data (which could include traffic and environmental conditions data) and generates a series of waypoints. These waypoints are used as inputs by the path planning layer, whose goal is to find a feasible trajectory to some target vehicle position and orientation, which could be defined by the next waypoint along the route. To achieve this, the path planning layer needs information on the current vehicle position and orientation from localization algorithms. It also requires a description of the environment from perception algorithms, with information such as

FIGURE 7.1 Task flow in the decision and system control process for CAVs.

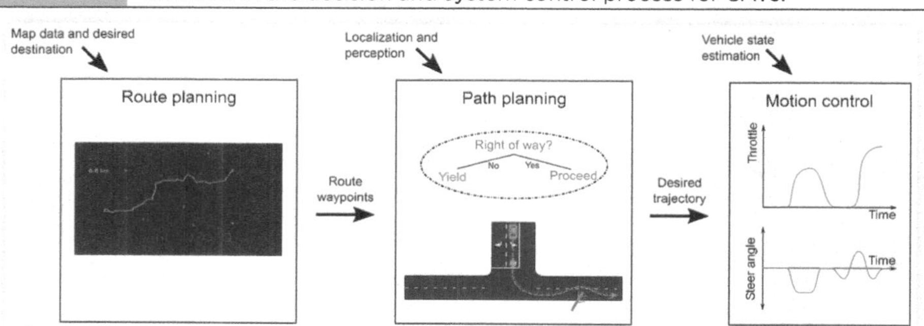

Left: Maxger/Shutterstock.com Center and Right: © SAE International.

the locations of the roadway and lanes, any traffic control devices, and the locations and predicted trajectories of other road users. The path planning layer processes this information through a set of behavioral rules and physical constraints to generate a desired trajectory. The motion control task aims to follow the desired trajectory given the current vehicle state.

This chapter also briefly discusses *motion control*, which is the execution of longitudinal (acceleration and braking) and lateral (steering) control maneuvers to follow the desired reference trajectory as identified by the path planning layer. Motion controllers constantly compare the current vehicle state (position, orientation, and velocity) with the reference, minimizing the difference between the two and accounting for factors such as steering and throttle delay, tire friction, and the vehicle's inertial properties. Path planning algorithms often consider the feasibility of maneuvers, and thus the limitations of motion control are a key part of path planning. However, the vehicle models employed in the motion control layer are often too complex to include explicitly in path planning.

Figure 7.1 shows the decision and control hierarchy as sequential, from left to right, but the route planning and path planning layers run continuously to update the reference route and path in response to changing road environments, or if the difference between the reference trajectory and the current vehicle state in the motion control layer grows very large. For example, consider the scenario presented in Figure 7.2. The CAV in red is traveling northbound (up) and intends to make a left turn at an intersection controlled by a four-way stop. The behavioral layer of the path planner prescribes that the subject vehicle must come to a stop at the stop bar before continuing onward. It arrives at the intersection before the other vehicles, and thus the path planner has determined that the subject vehicle has priority and proceeds to initiate a left turn, following a reference trajectory shown by the dashed line. However, a pedestrian has started to cross the intersection, conflicting with the CAV's intended trajectory, shown by the red arrow. The path planner must recompute the trajectory to avoid pedestrian conflict. The simple scenario depicted here highlights a few key path planning considerations; behavioral decision-making that accounts for traffic laws at a four-way stop, reference trajectories to predictably navigate intersection environments, and trajectory planning to account for changing environments.

FIGURE 7.2 Left-turn scenario depiction.

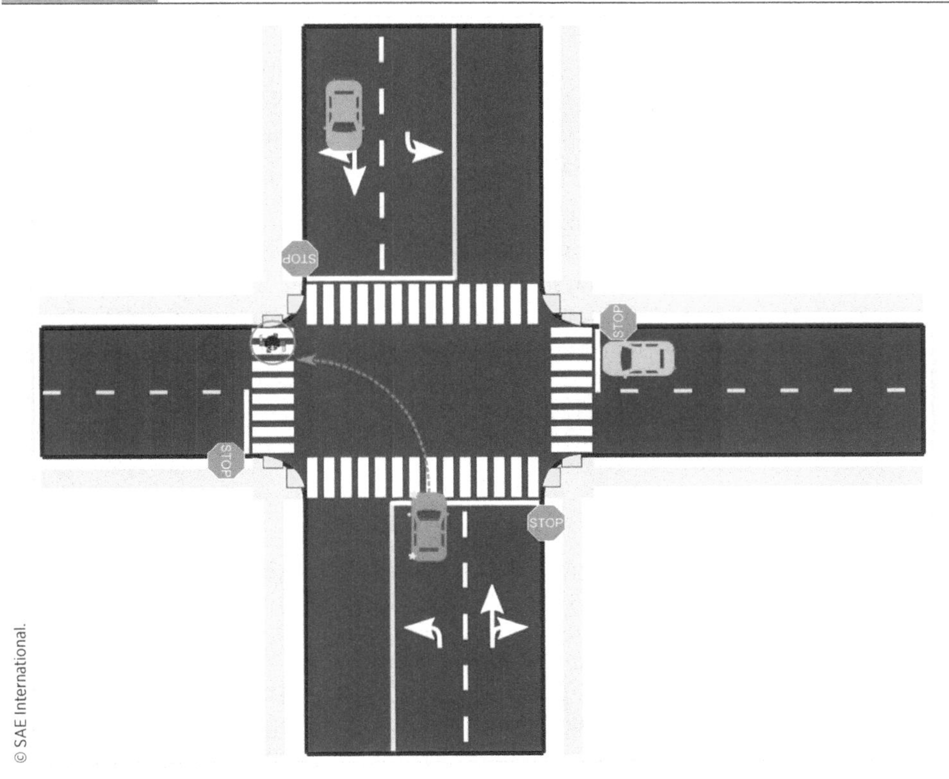

Path Planning Objectives

The primary goal of the path planning task is to define a feasible path or trajectory from the current state to some future state. A path description only includes information about *where* to traverse the environment, without prescribing the velocity or other time-based information. On the other hand, a *trajectory* includes information about *where* and *when* to traverse the environment. Algorithms may separate the path and trajectory planning problems, but ultimately the motion planning layer requires a reference trajectory to follow. Therefore, we will generally consider the output of the path planning task as a reference trajectory.

Both time and space are continuous, and there will, thus, always theoretically be an infinite number of feasible trajectories that achieve the path planning goal. However, feasibility is not the only consideration—the path planning problem is formulated to balance several objectives:

1. Safety: The trajectory should not cause a collision or violate the safety envelopes of other road users.
2. Comfort: The trajectory should be smooth, with minimal discomfort to vehicle occupants.

3. Speed: The trajectory should achieve the target state in the minimum time.
4. Efficiency: The trajectory should achieve the target state with minimum energy use.
5. Legality: The trajectory should not violate any traffic laws.
6. Behavioral Ethics: The trajectory should comply with a framework of subjective social rules on acceptable behaviors.

Not all the objectives are given equal weights. For example, in emergency situations, the path planner would be expected to generate an emergency maneuver that could cause some discomfort to the passengers in order to avoid a collision. Legality and behavioral ethics considerations can vary depending on factors, including whether the route definition includes structured or unstructured environments. The vehicle dynamics of a CAV must also be considered as a constraint, and planned vehicle motions cannot be unrealistic.

Structured Environments

Structured environments are those in which the traffic flow is well defined, and there exists a set of constraints to simplify the path planning problem. Structured environments generally have established guidelines for expected behavior in the form of traffic laws or driving manuals (Kolski, Ferguson, Bellino, & Siegwart, 2006). For example, in the intersection presented in Figure 7.3, conventional driving behavior would dictate that traffic coming from the south intending to turn right (east) travel in the right lane and come to a stop behind the stop bar, wait for a gap in eastbound traffic, and initiate a right turn when clear.

The preferred behavior depends on the conditions and sequence of events that transpire. In the three-way intersection example, the behavioral algorithm knows from the output of the route planning layer that reaching the next waypoint along the route requires a right turn at the intersection. Approaching the intersection triggers the algorithm to remain in or change to the right lane. Upon approaching the stop bar, the behavioral layer will dictate that the vehicle stops and waits for a gap in traffic. Once the perception system detects a suitable gap, a decision will be made to initiate a right turn. Though there are various approaches to behavioral decision-making, the predictability inherent to structured environments can simplify this task.

Structured environments also offer reference paths for vehicle motion. These reference paths generally follow basic guidelines such as following the center of a lane or tracing a smooth turn in an intersection. Example reference paths in a three-way intersection are shown in Figure 7.3. In practice, reference paths may be generated from HD map data or manually prescribed. The increasing availability of HD maps will further facilitate driving in structured environments. However, even when conditions allow a reference *path* to be followed without deviation, the desired *trajectory* that includes temporal information must still be computed, as the speed at which the path is to be followed may depend on factors such as traffic, weather, and energy use considerations.

FIGURE 7.3 Examples of reference paths in a three-way intersection.

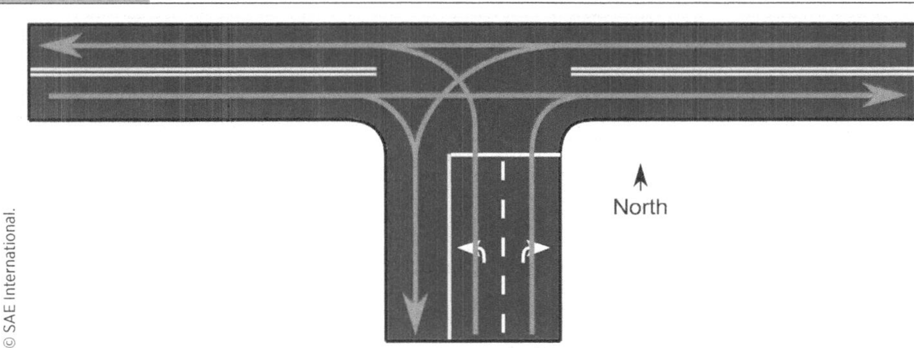

Deviations from Reference Paths

Following reference paths in structured environments has a few important benefits. First, it greatly simplifies behavioral decision-making and trajectory planning by reducing the range of possible motions. Second, it makes the vehicle trajectory more predictable, which increases the perception of safety for human occupants and other road users. Last, it allows for infrastructure-based systems and sensors to better track, analyze, and predict vehicle motion on the roadway to achieve network-level transportation goals such as reduced congestion and increased efficiency.

Even in highly structured environments, conditions may require deviations from established reference paths. For instance, the reference path may not be obstacle free, necessitating a preemptive or emergency maneuver to avoid the obstacle depending on when the obstacle is perceived. These two situations are presented in Figure 7.4. In the first example, the subject vehicle is driving along a four-lane roadway in the right lane and perceives a branch blocking the lane ahead well in advance, but that the adjacent lane is not blocked. As a human driver would be expected to do in such a circumstance based on established driving convention, the behavioral algorithm directs the CAV to perform a lane-change maneuver to the unoccupied passing lane that is adjacent.

In the second example, however, the obstruction is not perceived and classified sufficiently in advance, and the behavioral layer directs the CAV to search for and execute an emergency evasive maneuver. The selected maneuver must balance the various objectives according to their relative importance. In this instance, an emergency braking maneuver would not violate traffic laws and would allow the vehicle to stop before colliding with the branch. However, this maneuver could violate the fundamental safety objective if the vehicle stopped too close to the tree branch. On the other hand, a steering maneuver into oncoming traffic would avoid a collision but would violate traffic laws. A reasonable decision in the depicted circumstance would be to allow for traffic law violations and perform a steering maneuver. While intuitive to a human driver, this decision-making structure must be formalized by applying different weights and constraints to the various objectives. The path planner must consider the feasibility of various maneuvers based on the limitations

FIGURE 7.4 Examples of reference paths (dashed lines) and possible deviations (solid segments) to avoid a fallen tree branch blocking the current lane of travel.

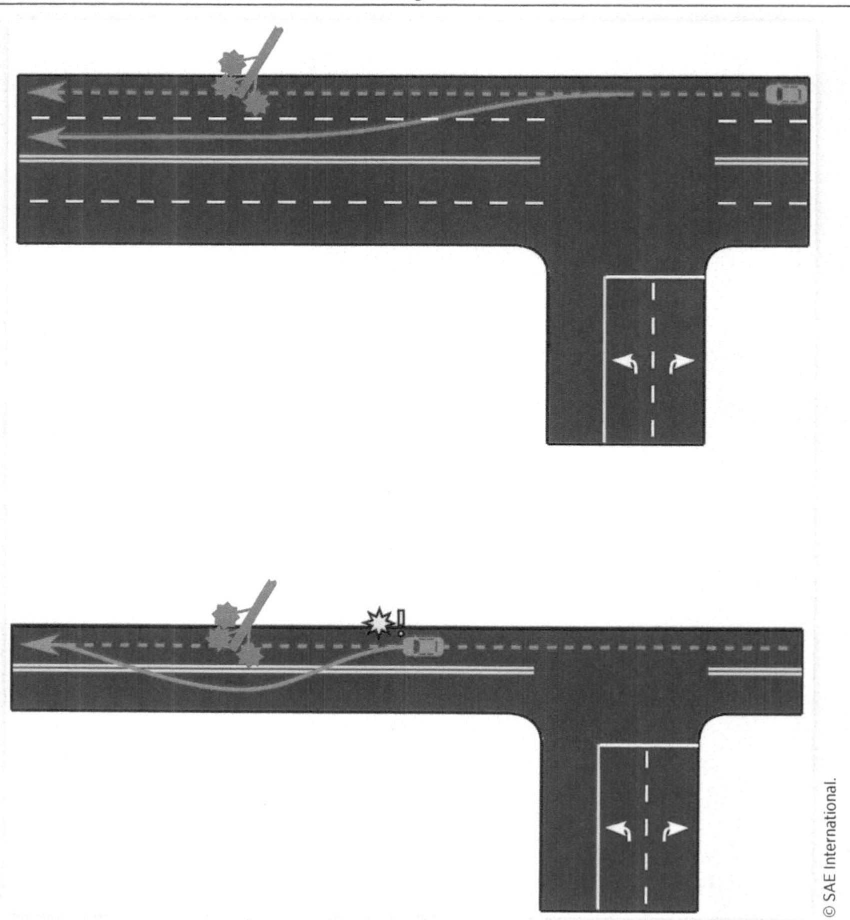

of vehicle dynamics and vehicle control authority. Traffic in the incoming lane would further complicate the task and introduce ethical considerations associated with the effect of the subject vehicle's decisions on other road users. These aspects introduce challenges to both behavioral and trajectory planning and will be discussed in further detail later in this chapter.

Unstructured Environments

The vehicle may sometimes operate in environments that do not have established guidelines for behaviors or preferred reference paths. Such environments are classified as *unstructured*. Examples may include construction sites and parking lots. The lack of reference paths means that the path planner generally encounters fewer behavioral constraints, but these environments may change in real time, requiring the constant calculation of updated paths (Lima, 2018).

For example, many parking lots do not have well-defined parking spaces or navigation paths. Nevertheless, the vehicle may be tasked with searching for and finding an open parking spot and then coming to a stop at this location at a specific orientation. Parking lots change frequently; possible objective positions may open up or become occupied by other vehicles leaving or entering parking spaces. In structured environments, it is often sufficient to employ local planners with short horizons to track a reference path or avoid an obstacle if required. In unstructured environments, a simple approach would entail using local planners to successively minimize the distance between the current position and the goal, but this carries risks. Consider the case shown in Figure 7.5. The vehicle, shown on the left, is attempting to reach the target, the small circle on the right. A simplistic approach would be to select an arc shown by the red curves that minimizes the straight-line distance between the current position and the goal. This strategy would cause the vehicle to proceed straight, directly toward the right of the image. However, it may get stuck behind the set of obstacles shown in black and be unable to execute a subsequent step. Thus, in unstructured environments, it is crucial to consider the feasibility of the entire path. In the example below, a global path search would track the sequence of points represented by vertices along the blue path from the current position to the goal (Kolski, Ferguson, Bellino, & Siegwart, 2006).

FIGURE 7.5 Unstructured environments showing the importance of planning the entire trajectory.

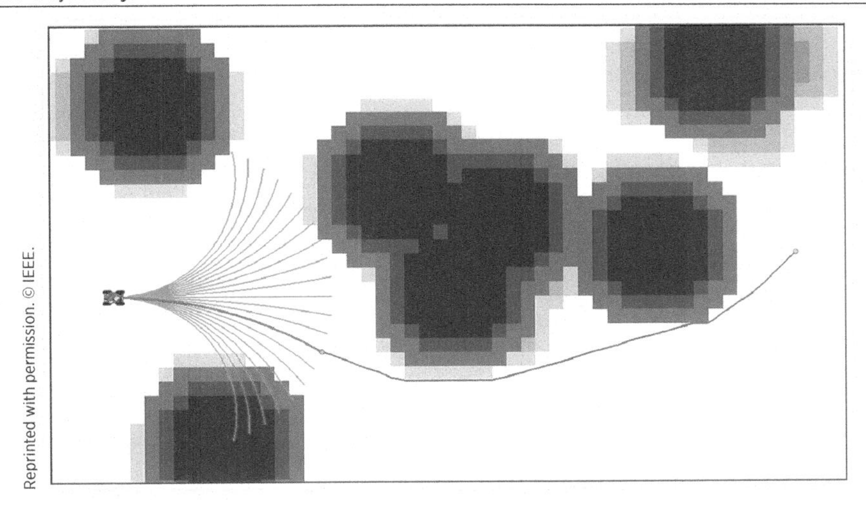

Behavioral Decision-Making

Before the vehicle can find a feasible trajectory through the environment, it must first establish what it wants to achieve. Ultimately, the maneuvers or actions selected would allow the vehicle to achieve the route-planning goals while remaining compliant with societal and legal guidelines governing reasonable behavior in the present situation. This involves evaluating the current environment and the movements of other road users and predicting their interactions with each other and with the subject vehicle.

There have been many methods proposed to achieve this complex and safety-critical task. Perhaps the most well-established method is the use of finite-state machines to switch among predefined behaviors depending on driving context. This was the predominant approach preferred by most competitors in the 2007 DARPA Urban Challenge. Other approaches that address some of the limitations of finite-state machines include probabilistic formulations and AI or learning-based methods.

Finite-State Machines

Finite-state machines offer a way to narrow down potential maneuvers based on the current driving context and established rules. Each state of the machine represents a type of driving behavior—for example, "turning left" or "merging into traffic." Transitions among states or behaviors are triggered by specific events. These may include the positions of other vehicles or objects, changes to environmental conditions, or the location of the vehicle along the desired route. In the 2007 DARPA Urban Challenge, many teams, including the winning vehicle from CMU shown in Figure 7.6, opted to construct a behavioral algorithm consisting of hierarchical finite-state machines (Urmson, et al., 2008; Kammel, et al., 2008; Bacha, et al., 2008). These implementations are hierarchical in nature because a state may contain several sub-states. For instance, a state that corresponds to "navigating an intersection" may prescribe some general driving behaviors and some that are specific to sub-states such as "approaching a 4-way stop" or "approaching an uncontrolled intersection."

FIGURE 7.6 CMU's winning entry to the 2007 DARPA Urban Challenge, "Boss," employed a finite-state machine driving logic.

Figure 7.7 presents an example of a hierarchical finite-state machine implementation used by the Carnegie Mellon team in the DARPA Urban Challenge. Inputs from the route planning algorithm and sensor fusion are used to determine the current situation and select an appropriate goal. Based on the current context (whether the vehicle is driving in its lane

FIGURE 7.7 Flow diagram showing an example of a behavioral algorithm composed of hierarchical state machines, used in CMU's vehicle "Boss" in the 2007 DARPA Urban Challenge.

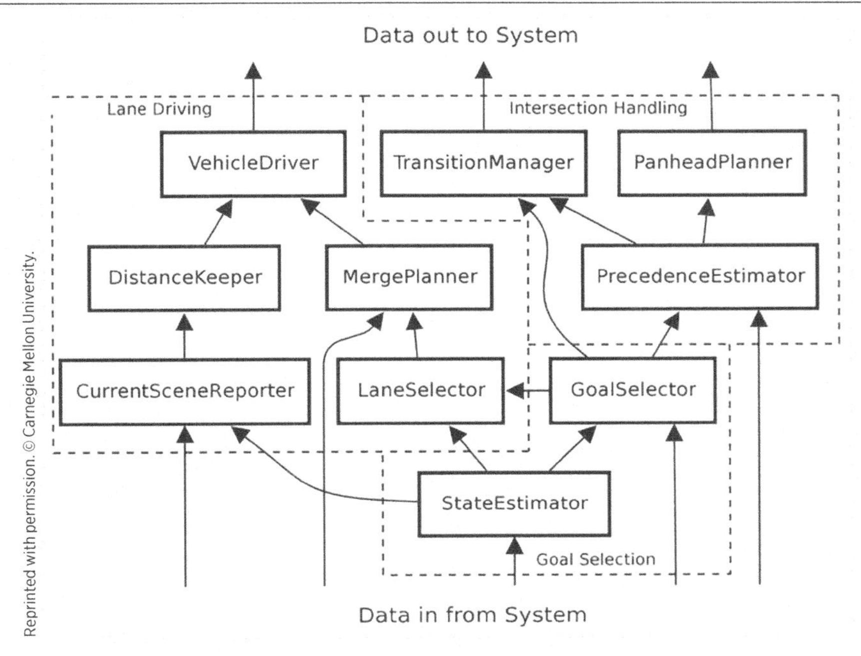

or traversing an intersection), an appropriate set of sub-behaviors is activated. The outputs of these are integrated and used by a trajectory planner to generate a feasible target trajectory through the environment.

Hierarchical finite-state machine implementations for behavioral decision have some advantages. They are simple and well defined, allowing for predictable behavior in scenarios that the state machine has been programmed to recognize. Each behavior can be tested and evaluated independently or semi-independently, allowing for more straightforward development and debugging. If some behaviors malfunction, others may still be able to operate and allow the vehicle to navigate the environment, though with reduced decision-making capability. While the rule-based formulation of state machines can work well in many circumstances and allow for strict adherence to traffic rules, it does not deal well with unfamiliar scenarios or with the uncertainties and complex interactions with other road users. In such situations, state-machine-based formulations may act excessively defensively and could completely freeze if none of the programmed states are able to describe a suitable behavior in the current situation (Schwarting, Alonso-Mora, & Rus, 2018).

Probabilistic Methods

In real-world driving, there is constant interaction among road users, with the result that common socially conscious behaviors may supersede traffic rules and guidelines in certain situations (Sadigh, Landolfi, Sastry, Seshia, & Dragan, 2018). For instance, if Vehicle A is turning out of a driveway onto a busy main road, it would be expected to yield as traffic on the main road has the right of way. However, a courteous driver in Vehicle B on this road may slow down a little to create a large enough gap with the vehicle in front to allow Vehicle A to safely turn right and merge into traffic. A human driver of Vehicle A should know to interpret Vehicle B's behavior as an implicit permission to initiate a turn. This knowledge may arise from past experience in similar circumstances or even direct communication with the driver of Vehicle B in the form of a nod or a wave. If Vehicle A were a CAV, it would need some similar way to estimate the likelihood that Vehicle B is indeed yielding to it. Figure 7.8 shows the examples of merging and overtaking vehicles. The top vehicle is taking an exit lane off the freeway, so its presence has a decreasing impact on the vehicles in the three lanes going to the left still on the freeway. The lead vehicle in the right-most lane impacts that vehicle behind it if it brakes while the lone vehicle in the left-most lane has an impact on the behavior of the other two (i.e., it could be unsafe for these two vehicles to change lanes directly into the left-most lane because of its presence), but the impact is lower because there is a lane of separation. Conversely, in the two lanes going to the right, the lone car overtaking the lead car in the right lane impacts the behavior of the latter directly and significantly.

FIGURE 7.8 The behavior of road users is not independent. Common actions such as merging and overtaking have an effect on the behavior of the surrounding vehicles.

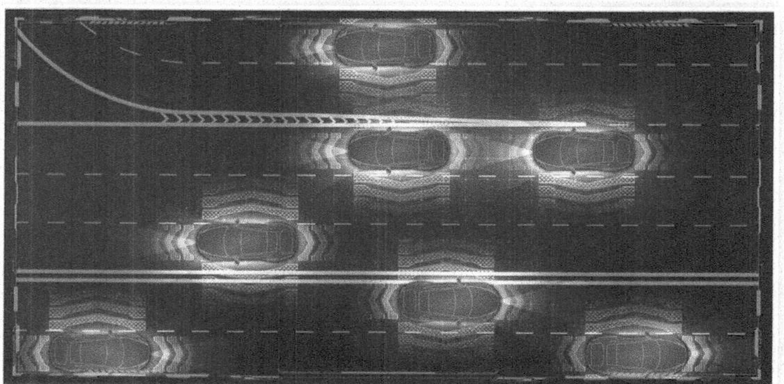

ZinetroN/Shutterstock.com.

Several methods have been proposed to try and account for the variable and unpredictable behavior of other road users, such as assuming other vehicles' trajectories have bounded random disturbances or considering a set of possible trajectories for other vehicles. However, these do not necessarily account for the effect that the subject vehicle has on the behavior of the other road users. Executing common behaviors such as lane changes, merging, or overtaking may cause other vehicles to change their behaviors in response (Galceran, Cunningham, Eustice, & Olson, 2015).

Probabilistic methods address this by employing models to predict the likelihood of various outcomes arising from the interactions among road users. One way to approach this is by employing what is often called a partially observable Markov decision process (MDP). MDPs provide a mathematical model for an environment with probabilistic behavior. In an MDP, the environment has a set S of individual states s. In the context of automated driving, a state s may include the behaviors of the subject vehicle and other road users. There is also a set A_s of available actions a_s in the state s. $P_a(s, s')$ is the probability that action a_s will cause a transition in the state from s to s'. $R_a(s, s')$ is the immediate reward for the same transition. The objective is to find a policy $\pi(s)$ that describes the actions of the driving agent for s.

CAV decision-making is constructed as a *partially observable* Markov decision process (POMDP) because the behaviors and states of other road users are not necessarily directly observable. While connectivity among CAVs can increase the observability of other vehicles' intentions, those of road users such as cyclists and pedestrians may be communicated in a connected vehicle environment. POMDPs of the type needed to model interactions among road users can be very computationally expensive and can be impractical to implement in real time when split-second decisions are needed. To address this, various approaches have been presented to try and reduce the complexity by introducing some prior knowledge about common road user behaviors.

The behavior of other road users may also be modeled using game theory. In this framework, each road user selects the behaviors that maximize some sort of benefit described by some reward function. As such, it shares some similarities with the POMDP formulation. Game-theoretic approaches try to predict likely responses of other road users to the subject vehicle's reactions, and the subject vehicle's reactions to other road users. The subject vehicle's ultimate reward depends on these interactions. As with all game-theoretic approaches, this type of formulation requires careful modeling of interactions and suitable reward functions to avoid the selection of unreasonably selfish or selfless behaviors. As the number of vehicles increases, the number of possible interactions grows exponentially. There is also the challenge of identifying a suitable reward. Unlike in, say, a game of chess where each piece captured from the opponent has a well-defined and instantly quantifiable reward, rewards in the "game" of driving are less obvious. To keep the mathematical problem tractable, specific driving scenarios can be treated in isolation and several simplifying assumptions may be made—for instance, that the behavior of a vehicle on a congested freeway is dominated by the behavior of the vehicle directly in front of it, ignoring interactions with most other vehicles on the road.

Learning-Based Methods

The methods described in the prior sections may work well when a good model of probabilities and interactions among road users is available, such as in highly structured freeway driving. These methods are less effective as the complexity and variability of the environments increase. Learning-based methods offer a way to define a set of policies without the need for explicit models of probabilities or interactions among road users. Instead, decision-making triggers are trained by examining large quantities of data and employing machine learning algorithms. Learning-based methods attempt to determine the appropriate action based on a large number of "features," which could include environmental conditions, roadway type, the location and trajectory of other road users, and the vehicle state. For a given combination of features, the algorithm learns to react in a certain way. Training the appropriate behavior is nontrivial and generally requires large quantities of data given the wide range of conditions and behaviors encountered in the real world (Schwarting, Alonso-Mora, & Rus, 2018).

Learning-based methods employed in CAV decision-making can be broadly categorized into *reinforcement learning* and *imitation learning* approaches. In reinforcement learning, the environment is essentially modeled as a traditional MDP with a specified reward function. The vehicle iteratively learns which kinds of actions increase the reward and which actions reduce it. As with the POMDP, the ultimate goal is to establish a set of behaviors that maximizes the reward in an environment where not all states may be observable. However, POMDPs with explicit mathematical model are amenable to optimization techniques to solve for a decision policy $\pi(s)$. On the other hand, reinforcement learning methods learn from past attempts to guide the algorithm toward a set of beneficial policies. There is therefore no guarantee of optimality in reinforcement learning, and the learned behavior isn't necessarily the one that truly maximizes the reward. However, because no explicit model of the decision process is necessary, reinforcement learning can be applied when the problem grows too large for probabilistic methods like POMDPs to be effective. Reinforcement learning algorithms require careful consideration and development to strike a balance between selecting behaviors that have already been learned and *known* to generate

FIGURE 7.9 High-fidelity simulation environments such as NVIDIA DRIVE Sim can be used to train and learn driving behaviors.

Photo courtesy of NVIDIA

a high reward and exploring the *unknown* decision space to try and find behaviors that may provide an even higher reward. Specifying the reward function is also non-trivial as unintended consequences could occur if not structured appropriately.

Reinforcement learning methods for CAV decision-making require the vehicle to navigate the environment many times to train the appropriate actions. Training the system in the real world requires a significant amount of time and resources and could pose safety risks, especially in the initial stages before the algorithm has a large enough set of prior attempts to learn from. Thus simulation environments like the one shown in Figure 7.9 are often employed (see Chapter 8). Simulations can be accelerated to speeds limited only by computational power, which reduces the development time. To reflect real-world driving, simulation environments can be constructed from data captured in the real world, allowing the algorithm to learn from the behavior of actual road users.

It is often difficult or impossible to define an explicit reward function to be used in reinforcement learning for environments as complex as those encountered in real-world driving. To address this, imitation-based methods learn from the behavior of expert demonstrators—in this case, the experts are real human drivers. The most straightforward type of imitation learning is *behavior cloning*, which essentially learns a decision-making policy using supervised learning techniques to mimic the actions of a demonstrator in the same state. This approach is suitable when environments are predictable and repeatable, like driving around an empty racetrack. But because behavior cloning does not account for how the agent's actions in a current state induced the next state, it can easily break down when small errors cause the agent to encounter a state for which it has no expert demonstration for reference. While there are ways to improve performance in such situations, such as having an export demonstrator to provide feedback and new demonstrations when errors grow, this approach is not scalable to the complex, changing environments encountered in real-world driving.

In a framework known as *inverse reinforcement learning*, the reward function of the MDP environment is learned from observation. A general form of the reward function may be manually specified or formulated as a neural network (similar to the ones described in Chapter 5). If manually specified, the inverse reinforcement learning algorithm is employed to find the weights and coefficient of the reward function's terms. In more complex environments, the reward function may be expressed as a neural network and the inverse reinforcement learning algorithm employs a simulation environment to compare the performance of the current policy with that of an expert before updating the reward function for the next iteration.

Vehicle- and infrastructure-based sensors and cameras allow for the capture and processing of large quantities of data that may be employed to help facilitate imitation learning. For example, Tesla has employed a "shadow driver" as part of its imitation learning process. This shadow driver is an automated driving agent that predicts and plans behaviors and movements, even while a human driver is operating the vehicle. However, it does not actively take over control to execute the maneuvers it has planned. Instead, the intent is to validate the agent's decisions and identify circumstances that lead to errors between the agent's predicted behavior and the human driver's actions. The sensor signals and video footage collected during these errors may be uploaded to train the agent, or manually processed and analyzed by a human for future software development purposes. Such an approach avoids collecting and training on an unnecessarily large quantity of data for relatively straightforward driving situations and focusing on those that pose greater safety risks. While the "shadow mode" approach (illustrated in Figure 7.10) allows for constant monitoring and learning from expert demonstrations, it cannot account for events that may transpire if the actions selected by the automated driving agent were actually taken. For instance, if the agent decides to change lanes to overtake a slower moving vehicle, but the human driver instead slows down and remains behind the lead vehicle, data on the reactions and behaviors of other road users to the potential lane change would not become available for analysis. Hence, despite the availability of vast amounts of data for human-driven miles that can be employed in imitation learning, the use of simulation environments is an important part of automated driving agent development.

Learning-based methods are well suited to address many driving tasks in a combined manner and can sometimes blur the hierarchical system structure described in this chapter.

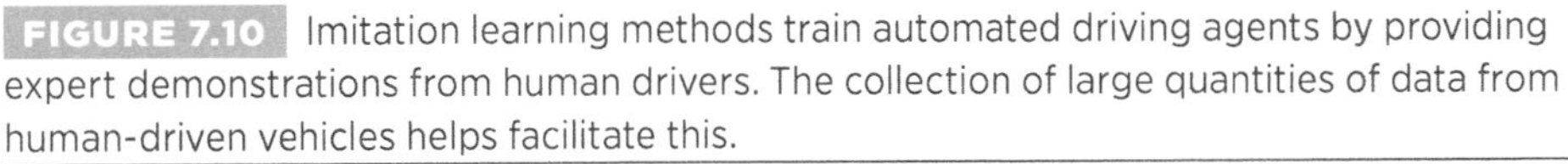

FIGURE 7.10 Imitation learning methods train automated driving agents by providing expert demonstrations from human drivers. The collection of large quantities of data from human-driven vehicles helps facilitate this.

Since they do not need to rely on manual definitions of driving states or an explicit scenario-dependent problem formulation, these methods can be applied to *end-to-end* motion planning. Unlike the hierarchical approach described in much of this chapter, end-to-end motion planning seeks to map camera sensor inputs directly to steering/throttle/braking controls. End-to-end planning is discussed further in a later part of this chapter.

While learning-based behavioral decision-making methods can be empirically shown to result in acceptably safe behaviors in a set of scenarios, these approaches result in neural networks that are difficult to interpret by humans, which makes it difficult to independently verify their robustness, known as the AI black box problem. Progress has been made recently in understanding how path planning decisions based on AI algorithms are made, but unless and until full transparency is available, this technique can be problematic.

Behavioral Ethics

Ethics in path planning and motion control is an evolving topic that incorporates ideas from behavioral psychology, sociology, and engineering. Humans constantly make ethical judgments when driving. These judgments could be as trivial as whether it is permissible to cross a solid line on the inside of a curve when turning or when passing an obstacle on the road, or as dire as where to steer a truck whose brakes are perceived to have failed. The ethical frameworks guiding CAV decision-making must consider the effects various actions will have on other entities in the environment.

Two broad classifications of ethical frameworks are deontological ethics and consequential ethics. Deontology posits that the morality of any action depends on whether it adhered to some set of rules, while consequentialism evaluates the morality of a decision based on its consequences. Most ADS incorporate both types of ethical frameworks in their algorithms (Thornton, 2018).

Decision-making based on a deontological ethical framework is constrained by a certain set of rules. The finite-state machines discussed in the previous section can be thought of as a type of deontological framework where constrained driving states are defined by the vehicle state and environmental conditions. Switching among states is triggered by certain events. Behavioral decision-making in CAVs that relies on exceedingly strict deontology runs the risk of causing the vehicle to become stuck with no allowable path forward to the next route waypoint. To address this, hierarchies may be established that allow violation of certain lower-level rules while adhering to higher-level rules. Because deontological automated driving can be specified by a clear set of priorities, the reasoning behind programming decisions is clear and easily auditable.

Consequential ethics, on the other hand, evaluates the morality of a decision only based on its consequences or costs. An action can be justified if the outcome or expected outcome is generally positive. Of course, characterizing what qualifies as a positive outcome is challenging and fraught with ethical issues itself. Automated driving algorithms often employ optimization routines that seek to determine an action or maneuver that minimizes certain costs. A more comprehensive discussion of constraints and costs, analogous to deontological and consequentialist considerations, respectively, is offered in later sections of this chapter in the context of optimization-based trajectory planning.

A simple example contrasting the two ethical frameworks is illustrated by the scenario depicted in Figure 7.11, where a vehicle encounters an obstacle in its path. There is a path around the obstacle that does not require the vehicle to violate the double yellow line constraint separating travel lanes in opposing directions. A behavioral decision-making algorithm that treats traffic rules deontologically as rigid constraints would be stuck.

FIGURE 7.11 An automated vehicle approaches an obstacle, a fallen tree branch, in its path. To proceed, it must violate the double yellow line separating oncoming lanes of traffic. An algorithm may respond differently depending on whether it treats driving rules more according to a deontological (A) or consequentialist (B) framework for ethical decision-making.

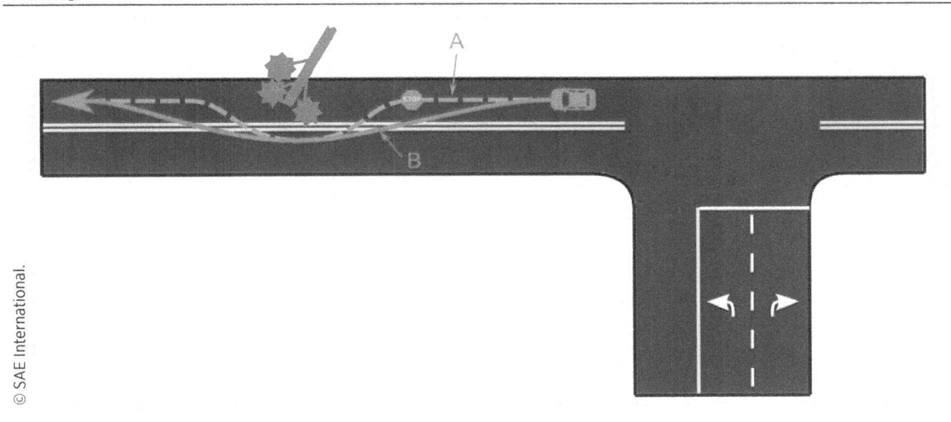

A hierarchical set of rules could be established that allows the vehicle to cross the double yellow line after coming to a stop and observing no other road users in the opposing lane. This is depicted by the path marked "A." Meanwhile, an algorithm more closely aligned with consequentialism may assign a cost to crossing the double yellow line without any specific constraint associated with it. Thus the vehicle would cross the yellow line in order to avoid hitting the obstacle without any rule-based permission structure explicitly allowing it to do so. To avoid hitting the obstacle, the cost associated with the collision must be sufficiently high or treated as a deontological constraint.

In automated driving, both the deontological and consequentialist ethical frameworks can be designed to operate according to a set of social and moral preferences. If a vehicle cannot proceed without violating certain rules, the algorithm can allow constraints to be relaxed according to a predefined hierarchy of driving rules. From a consequentialist viewpoint, potential outcomes can be assigned different costs based on a judgment of their impact on society.

Moral Dilemmas

Establishing a hierarchy of driving rules or relative costs of different projected outcomes can easily become fraught with moral dilemmas, especially with elevated stakes or life-or-death decisions. Whether a decision is considered moral depends on a variety of social and cultural factors that cannot be easily expressed in a comprehensive manner.

The trolley problem is a well-known thought experiment that incorporates these sorts of moral dilemmas, albeit with a rather morbid framing. There are a number of versions and variants of the trolley problem, but a basic formulation is as follows:

> You observe a runaway trolley that appears to be without functioning brakes. In its path are five people who are likely to be killed if the trolley continues on its path. You are next to a switch that, if you pressed it, would divert the trolley to a side track and avoid the five people. However, there is a person standing on the side track who is certain to be killed if the switch is pressed. What do you do?

A pure consequentialist may profess that it is preferable to press the switch because the outcome of one certain death is preferable to five likely deaths. A deontologist who adheres to a strict rule to never kill someone may profess that it is better to do nothing and passively risk the five deaths than to actively kill one person. Researchers have found that the people's decisions on what would be the moral choice are very sensitive to how the questions are framed and to any additional details offered about the six fictional people in the thought experiment. This type of thought experiment has been extended to automated driving scenarios. Perhaps the most famous is the Moral Machine from the MIT Media Lab (Awad, et al., 2018).

The Moral Machine Project

The Moral Machine experiment, active from 2016 to 2020, was created by researchers at the MIT Media Lab to collect perspectives on the ethics of various decisions made by CAVs. The experiment generated scenarios in which the vehicle must decide between the "lesser of two evils" and used a web interface to ask participants which option was more acceptable to them. An example is shown in Figure 7.12, where a vehicle with five occupants is approaching a crosswalk with five pedestrians. The participants are asked "*What should the self-driving car do?*" and prompted to select one of the two presented options. In the first, the car steers to avoid the pedestrians but strikes a barrier, killing the five occupants. In the second, the vehicle continues straight ahead, striking and killing the five pedestrians in the crosswalk.

FIGURE 7.12 An example of a moral dilemma presented in the Moral Machine experiment. The study participant is presented with a moral dilemma such as the one pictured and must decide which represents the preferable decision.

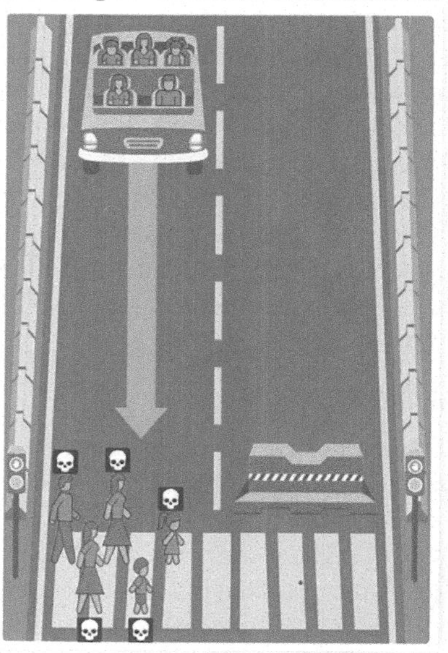

Over the course of the Moral Machine project, 40 million decisions were recorded from participants in 233 countries and territories. While some decision trends were global, the outcomes showed variations consistent with known cultural and economic variations in ethical judgments. This indicates that the algorithms underlying CAV decision-making may need to be tuned to account for variations in expected moral behavior in different vehicle markets. Since CAVs may represent the most significant example of machines making life-or-death decisions, this itself poses ethical questions regarding what extent developers should replace their own ethical preferences with those more prevalent in other regions of the world.

Some key trends observed in the data included differences between so-called individualistic cultures and collectivistic cultures (Hao, 2018; Awad, et al., 2018). Data from countries commonly identified as having a prevailing collectivistic culture, many of which are geographically in East Asia, showed a preference toward sparing the elderly. The researchers hypothesized that this was perhaps a result of the respect that is due to older members in these societies. A similar grouping of geographic regions was observed in how much they emphasized sparing a greater number of lives over other considerations. Countries identified as having a more individualistic culture tended to indicate a preference toward sparing a greater number of lives, while those in collectivistic cultures placed a relatively smaller emphasis on the number of people in harm's way in the presented scenarios. Some trends were not entirely consistent with the hypothesized distinction between collectivistic and individualistic cultures. For instance, Japan and China, both considered as more culturally collectivistic, generally showed opposite preferences with respect to sparing pedestrians over vehicle occupants. Respondents in Japan were more likely than average to spare pedestrians over vehicle occupants, while those in China were more likely than average to spare vehicle occupants in the same scenarios.

Thought experiments are a good way to understand people's values and motivations for decision making, and the Moral Machine has initiated and contributed to many meaningful discussions on CAV ethics. However, the Moral Machine project has some notable limitations. The participants must have had internet access to participate and were self-selected. They also were better educated and younger than the general population. Further, the scenarios presented in the Moral Machine were deliberately designed to be high-stakes, life-or-death decisions with inevitable casualties and inherent moral quandaries. Such situations are exceedingly rare, and the trolley problem on which the Moral Machine is based is itself is a contrived scenario with explicit outcomes from decisions that are not replicated in real-world driving, where there is considerable uncertainty in outcomes based on chosen actions. There is a risk that overemphasizing such situations will take focus away from developing safe, robust decision-making algorithms that offer improved safety over today's human-driven vehicles.

Regulatory Guidance

To date, there has been limited regulatory or governmental guidance for CAV ethics. In 2017, an Ethics Commission established by the German Ministry of Transport and Digital Infrastructure released a report offering a set of twenty ethical guidelines regarding both the regulation and function of CAVs (Di Fabio, et al., 2017). Some of the key points are summarized below:

- The primary purpose of automated transport systems is to improve safety for all road users.
- Accidents should be prevented wherever practically possible, and the technology should be designed such that critical situations should not arise in the first place.
- In hazardous situations that prove to be unavoidable, the protection of human life enjoys top priority over other legally protected interests.
- Genuine dilemmas cannot be standardized or programmed to the point that they are ethically unquestionable. In unavoidable accidents, distinction based on personal features is prohibited, and parties involved in generating mobility risks must not sacrifice non-involved parties.

In 2020, the European Commission published an independent expert report on the Ethics of Connected and Automated Vehicles (Horizon 2020 Commission Expert Group, 2020). It stated that CAVs should be expected to reduce physical harm to persons and should prevent unsafe use through inherently safe design. In the context of managing dilemmas, it noted:

> While it may be impossible to regulate the exact behaviour of CAVs in unavoidable crash situations, CAV behaviour may be considered ethical in these situations provided it emerges organically from a continuous statistical distribution of risk by the CAV in the pursuit of improved road safety and equality between categories of road users.

This statement describes rather than prescribes a methodology for ethical decision-making. As more regulatory bodies around the world begin to provide guidance on CAV safety, they should consider how this guidance can stimulate and encourage innovation. Given the complexity of real-time behavioral decision-making in CAVs and of defining ethical actions in real-world scenarios, regulatory efforts are likely to focus more on methodologies to ensure ethical behavior rather than defining these behaviors explicitly.

Trajectory Planning

In the hierarchical description of vehicle autonomy, trajectory planning occurs once the behavioral decision-making algorithm has established which action should take place. Usually, the trajectory planning task involves following pre-established reference paths in structured environments. However, the behaviors of other road users, the presence of obstacles traversing unstructured environments, or other factors may require a deviation from these reference paths. While the behavioral algorithm prescribes *what* maneuvers to execute, the trajectory planner describes *how* the maneuver should be executed and outputs a desired path through the environment along with speed and/or time information. There are, in theory, an infinite number of ways to execute any given maneuver. Consider again the scenario of a vehicle encountering a fallen tree branch, depicted in Figure 7.13. The vehicle shown in red is following a reference path, indicated by the dark dashed line, through the structured environment but must deviate from it to maneuver around the obstacle. The behavioral algorithm adopts a consequentialist approach to violating the double yellow line and, since the sensors do not perceive any no traffic in the oncoming lane, determines that it is safe and reasonable to maneuver around by briefly entering the oncoming traffic lane.

The desired trajectory around the obstacle is found by the trajectory planner, which considers many factors in its calculations. In Figure 7.13, the dashed trajectory marked “A” spends the least time in the opposing lane of traffic. This minimizes the amount of time spent in a condition that violates this basic traffic rule. However, it passes very close to the obstacle and results in the sharpest turns, which could cause discomfort to the vehicle occupants. Trajectory “B” has a smoother curve and imparts lower lateral force on the vehicle occupants, but it still passes the obstacle at a close distance. Trajectory “C” maintains a larger separation with the obstacle but spends the longest time in the opposing lane of traffic and deviates the most from the reference trajectory. This example illustrates that the trajectory planning process is sensitive to how much relative importance is placed on each path planning objective.

FIGURE 7.13 Three examples of trajectories that avoid the fallen tree branch. Each of the three trajectories prioritizes certain path planning objectives at the expense of others.

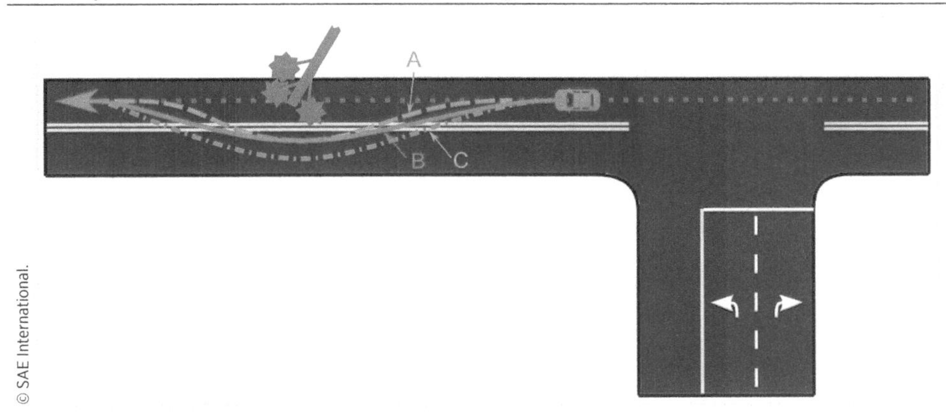

Two common approaches to trajectory planning are optimization-based methods and graph search and sampling methods. An overview of each is provided in this section.

The trajectory planning task can be described as an optimization problem that seeks to minimize some cost function while subject to a set of constraints (Paden, Čáp, Yong, Yershov, & Frazzoli, 2016):

$$\begin{gathered}\min_{q \in Q} \; J(q) \\ \text{subj. to} \\ q(t=0) = x_{initial} \\ q(t=T) \in X_{goal} \\ \Gamma_i(q(t)) = 0 \quad \forall t \in [0,T], \quad i \in 1 \ldots M\end{gathered} \tag{1}$$

In the above formulation, Q is the set of all possible trajectories and q is any single element of this set. $J(q)$ is the cost associated with following the trajectory q and may be designed to penalize (add costs to) characteristics such as large lateral accelerations, jerky motions, or deviations from reference trajectories. The optimization seeks to find the trajectory that minimizes the cost remaining compliant with a set of constraints. At the current time $t = 0$, the trajectory should start at the current state $x_{initial}$. At some future target time $t = T$, the trajectory should be such that the vehicle's final state is in some target region X_{goal}. The vehicle state definition usually includes the position, orientation, and velocity.

There is also a set of constraints $\Gamma_{i=1 \ldots M}$ that encode things such as obstacle avoidance, vehicle dynamics, and trajectory feasibility. Some factors can be considered either constraints or costs. For example, in Figure 7.13, it is reasonable to consider avoiding the fallen tree branch as a hard constraint. However, if violation of traffic rules were a hard constraint, there would be no feasible path around the obstacle as all possibilities require crossing the double yellow line. Rather than imposing a hard constraint on traffic rules, violations can be added to the cost function and assigned a very high cost—we would only want to violate traffic rules if absolutely necessary. Furthermore, to account for errors in perception and localization, we may want a trajectory that passes by the obstacle at a safe distance and add a cost to a trajectory that passes by the branch too closely.

Optimization-Based Methods

If formulated in a certain way, the trajectory planning problem can be solved directly by established continuous optimization techniques. To facilitate this, the constraints $\Gamma_i(q(t)) = 0$ are generally expressed as a set of equality and inequality constraints:

$$f_j(q(t)) = 0 \quad \forall t \in [0,T], \quad j \in 1\ldots N \tag{2a}$$

$$g_k(q(t)) \leq 0 \quad \forall t \in [0,T], \quad k \in 1\ldots P \tag{2b}$$

Depending on the type of solver being employed, it may be more efficient to formulate the problem as an unconstrained optimization. Then the equality and inequality constraints would be included in the cost function by using penalty or barrier functions, respectively.

Optimization-based methods can be constructed to explicitly account for the vehicle controls by parameterizing the trajectory by control inputs u, and the optimization problem can be expressed as follows:

$$\min_{u \in U} \quad J(u,x) \tag{3a}$$

$$\text{subj. to} \quad x(t=0) = x_{initial} \tag{3b}$$

$$x(t=T) \in X_{goal}$$
$$\dot{x}(t) = \phi(x(t), u(t)) \quad \forall t \in [0,T]$$
$$f_j(u(t), x(t)) = 0 \quad \forall t \in [0,T], \quad j \in 1\ldots N$$
$$g_k(u(t), x(t)) \leq 0 \quad \forall t \in [0,T], \quad k \in 1\ldots P$$

Now the objective is to find the set of control inputs that minimizes the cost function. We also have an expression for the vehicle dynamics and response to control input: $\dot{x}(t) = f(x(t), u(t))$. In practice, this continuous differential equation is discretized and expressed as a state transition: $x(t + 1) = \Phi(x(t), u(t))$. This formulation is known as model predictive control (MPC) since it employs a model of the vehicle dynamics and predictions of its response to generate a sequence of control inputs. MPC is well suited to motion control, as will be described later.

Optimization-based methods have a powerful mathematical description and can quickly find optimal or near-optimal trajectories under certain circumstances, especially when the range of possible trajectories is limited. However, because they can converge to local minima, they may ignore large parts of the feasible trajectory space in complex or unstructured environments. The choice of global optimization algorithm is thus important to avoid local minima.

Graph Search and Sampling Methods

To address the problem of trajectory planning in complex environments, sampling methods search through the space of globally feasible paths that do not cause a collision or other grave violation. This space is known as the *configuration space*. Each path is represented on a graph $G = (V, E)$, where V is a set of vertices and E is a set of edges. G should be constructed in such a way that all possible paths are feasible both from a safety perspective (do not intersect with obstacles) and physical perspective (the paths are feasible for the vehicle to follow). A simple configuration space could be analogous to the 2D planar surface that is free of obstacles, though orientation and velocity could be incorporated with added dimensionality. Since the configuration space contains an infinite number of possible graphs and paths, it is first discretized. If a full geometric description of the environment and obstacles is available, there are several established algorithms that can find minimum-distance feasible paths that are also limited to some maximum curvature. Often, however, a full geometric description of the environment is not available, and there are more constraints on any desired trajectory than a simple limit on maximum curvature.

Sampling methods can be employed to construct a graph by first discretizing and exploring the configuration space and identifying the set of reachable vehicle states without the need for a full geometric representation of the environment. These methods employ a strategy to project a set of path segments, or path primitives, forward from the vehicle's current state through maneuvers. One approach that has been well studied is the use of clothoids, which are curves whose curvature increases linearly with distance traveled. Clothoids, depicted in Figure 7.14, are also known as Euler spirals or Cornu spirals. They are commonly used in highway and railroad engineering to design transition curves. For drivers, this geometry translates to a constant rate of change of steering angle. For passengers, it is perceived as a smooth, comfortable change of lateral acceleration. The curves used for path segments can also be generated by observing and imitating the demonstrations of expert drivers. The segments are then joined to construct a discretized graph of the configuration space.

FIGURE 7.14 (a) An example of a clothoid, characterized by a smooth change in curvature. (b) Clothoids are commonly used as transition curves in roadway design, for instance, between a straight roadway and the tight maximum curvature of one of the loops in a cloverleaf interchange. (c) Examples of feasible future path segments that avoid the highlighted obstacles. Future path segments are constructed as clothoids with curvatures compatible with the current and prior path shown by the dashed curve.

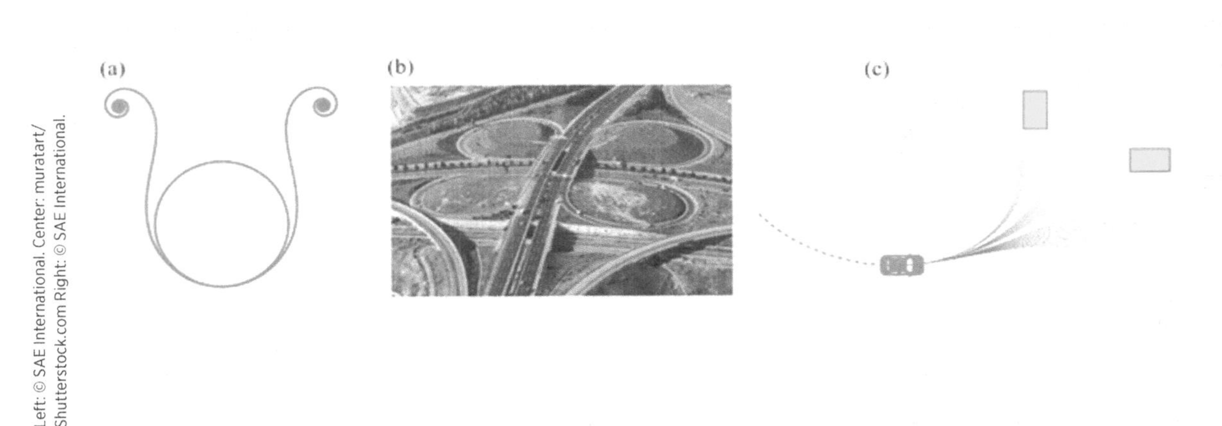

Since the graph is a discrete representation of a continuous space, there is always the possibility that the discretization process evolved in such a way that there exists no feasible path on the graph while one may exist in the real world. It is also possible that the optimal path on a discretized graph is much more expensive than the *true* optimal path in the continuous configuration space. There are many established approaches to searching through a graph for an optimal route, such as the A* and D* algorithms and their variants. Some of these variants incorporate modifications to try and find shortcuts on the graph between vertices that may not explicitly be joined by path primitive edges.

To overcome the computational and feasibility issues with constructing a full discretized graph of the configuration space and conducting a search over this graph, *incremental* search techniques attempt to search the configuration space by iteratively constructing paths that branch out from the current state. Subsequent steps of the branching process start from a random edge of the tree. Once the tree has grown to reach the target or goal region, a feasible path has been established. Rapidly exploring Random Tree (RRT) algorithms are a class of incremental planners that extend branches by selecting a random sample from the free configuration space and growing the tree in that direction. There are several RRT variants; RRT* offers optimality guarantees under certain conditions, while RRT^x allows for incremental replanning as there are changes to the environment.

Search and sampling methods and optimization-based methods can often operate together on the same vehicle. For instance, highly structured environments lend themselves well to optimization-based local trajectory planning, as the geometries tend to be well defined and there is a reduced possibility of converging to a suboptimal local minimum solution. The same vehicle may employ a tree search method in unstructured or congested urban environments that have many other road users or obstacles. The successful teams in the DARPA Urban Challenge used a combination of these approaches.

Motion Control

Once a desired path or trajectory has been established by the trajectory planner, the vehicle must be able to follow it, even in the presence of modeling errors, environmental disturbances, or other uncertainties. The objective of motion control algorithms is to generate the appropriate control inputs for trajectory tracking in a way that is robust to uncertainties. For instance, consider a very general dynamic vehicle model $\dot{x} = f(x,u)$, where x is the vehicle state and u is the control input, and a desired trajectory $x_{des}(t)$, the goal is to find the control u that results in an acceptably small error between $x(t)$ and $x_{des}(t)$ for the current and future time t.

Motion controllers commonly employ a single-track model of the vehicle, sometimes called the *bicycle model*. Figure 7.15 presents an illustration of this model. The distance from the vehicle's center of mass to the front and rear axles are l_f and l_r, respectively. There is a global reference frame with global x and y directions to define errors $\hat{e}_x$ and $\hat{e}_y$, respectively. The vehicle state x includes information on the position $\bar{p}$ of the vehicle's center of mass in the global reference frame, heading angle $\bar{\theta}$, planar velocity $\dot{\bar{p}}$, and angular velocity $\bar{\omega}$. The front wheel has a steer angle δ with respect to the vehicle heading. Both the front and rear tires have slip angle α_f and α_r, respectively, which is the difference between the vehicle heading and the actual direction of velocity of the front and rear tires. The model shown depicts the front and rear wheel angular velocity ω_f and ω_r, which is required to model longitudinal propulsive and drag forces due to tire/road interactions. Not all the elements of the single depicted track model are required for all motion control methods described in this chapter. For instance, kinematic path tracking controllers do not consider vehicle dynamics, and therefore do not consider lateral and longitudinal tire slip.

FIGURE 7.15 Graphical illustration of a single-track vehicle model. This version includes rotational degrees of freedom for the front and rear wheels to account for longitudinal tire forces.

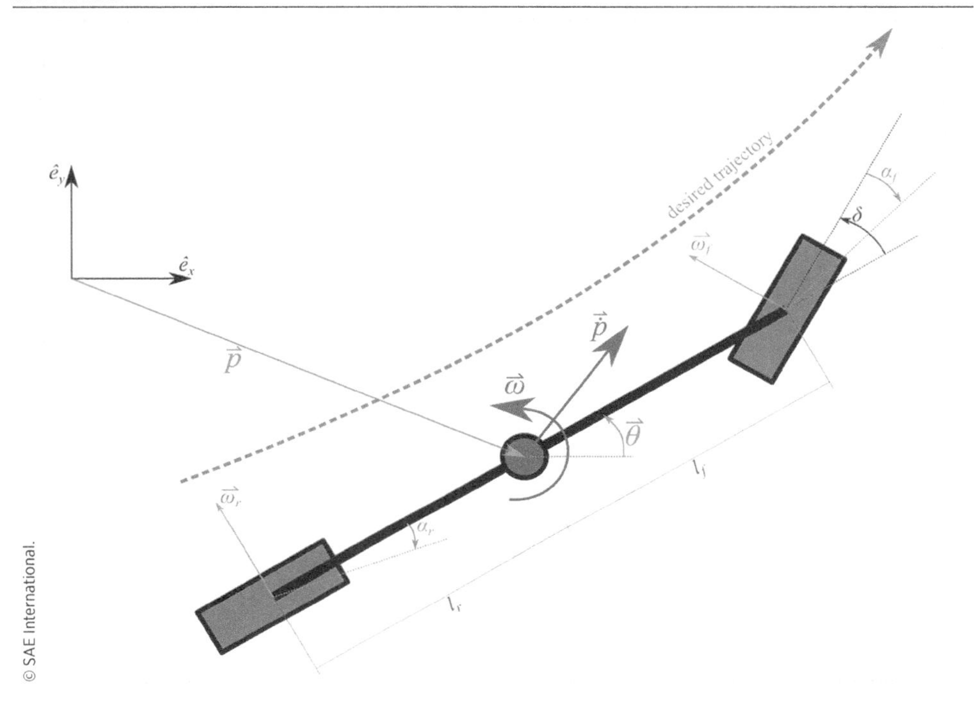

Kinematic Path Tracking

The objective of path tracking controllers is to follow a desired path in space. This contrasts with trajectory tracking controllers that seek to follow a desired trajectory in both time and space. Path tracking controllers only modulate the steering angle, while speed or throttle control is performed separately.

The pure pursuit controller is a fundamental, theoretically straightforward approach to path tracking. Pure pursuit is a look-ahead control strategy that picks a target point at some look-ahead distance l_t from a reference point on the vehicle. Most vehicles are front-wheel steer, and in kinematic models, the path of a vehicle is always tangential to its rear axle, so the center of the rear axle is a reasonable reference point to select. An arc is drawn from the rear axle to the target point and the front steering angle δ needed to trace this arc is calculated from the Ackermann steering model as

$$\delta = \left(\frac{2(l_f + l_r)}{l_t} \sin(\gamma) \right) \tag{4}$$

where γ is the angle between the current orientation of the vehicle and the target point. Figure 7.16 presents the geometry of the pure pursuit controller.

FIGURE 7.16 Illustration of the parameters in the pure pursuit path tracking controller. The controller calculates the steering angle δ required to reach a target point on the path a distance l_t from the vehicle.

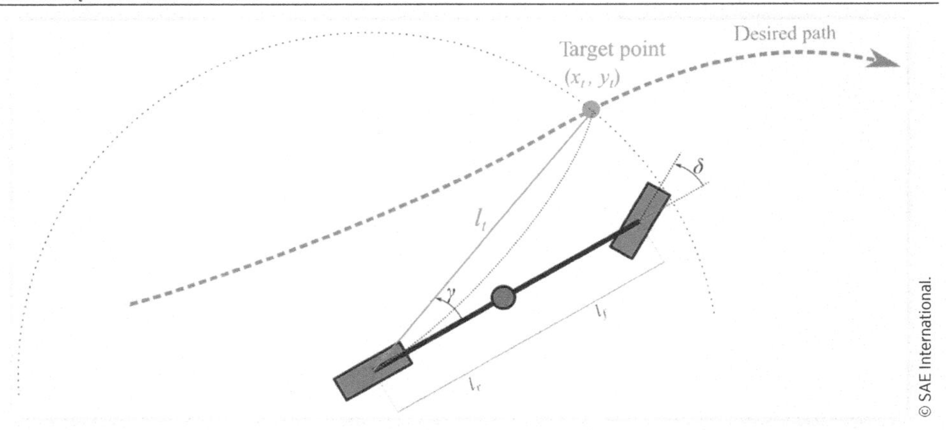

The only control parameter in basic path pursuit is the look-ahead distance l_t, which often increases with vehicle velocity to avoid jerky lateral motions at higher speed. Despite its simple and straightforward implementation, pure pursuit forms the basis for many path tracking controllers. It can be useful for steering control at low speeds where the vehicle motion is sufficiently well described by the kinematic model, such as in congested traffic or unstructured environments such as parking lots. It was employed by several teams in the DARPA Urban Challenge. However, pure pursuit has some important limitations. First, when the curvature of the desired path changes, there will always be some error between the vehicle's position and the desired path. The deviation can be reduced by selecting smaller values for l_t. However, if the distance between the desired path and the vehicle's current position is greater than, l_t, a target point will not be found, and the controller will not be defined.

Another set of path tracking controllers employs position feedback to minimize the lateral error between the vehicle's position desired path. A general illustration of the control parameters employed by these types of controllers is shown in Figure 7.17. The distance perpendicular from the path to a reference point on the vehicle is e_t. $\hat{t}$ is the direction tangent to the path, and θ_t is the difference in heading between the vehicle's orientation and $\hat{t}$. The controller computes the steering angle δ that keeps both e_t and θ_t small. To maintain generality, Figure 7.17 depicts the vehicle's center of gravity as the reference position from which to measure position error e_t. However, position tracking algorithms commonly employ the front or rear axles as reference points.

Trajectory Tracking

In contrast to path tracking, trajectory tracking involves following a trajectory in both space and time. Thus, at time t there is a specific desired position $\vec{p}_{des} = (x_{des}, y_{des})$, speed v_{des}, and angular velocity ω_{des}. To express the difference between the actual and desired trajectory, it is helpful to express this error in a vehicle-fixed reference frame, as shown in Figure 7.18. This allows for the separation of longitudinal and lateral position error x_e and y_e, respectively. θ_e is the error in heading angle. Kinematic trajectory tracking controllers seek to minimize error in x_e, y_e, and θ_e by varying speed v and angular velocity ω. The throttle/brake position and steering angle can then be back-calculated from v and ω.

FIGURE 7.17 General illustration of the parameters used in path position feedback control. The steer angle δ depends on the lateral distance to the path e_t and the difference θ_t between the vehicle heading and the path tangent $\hat{t}$.

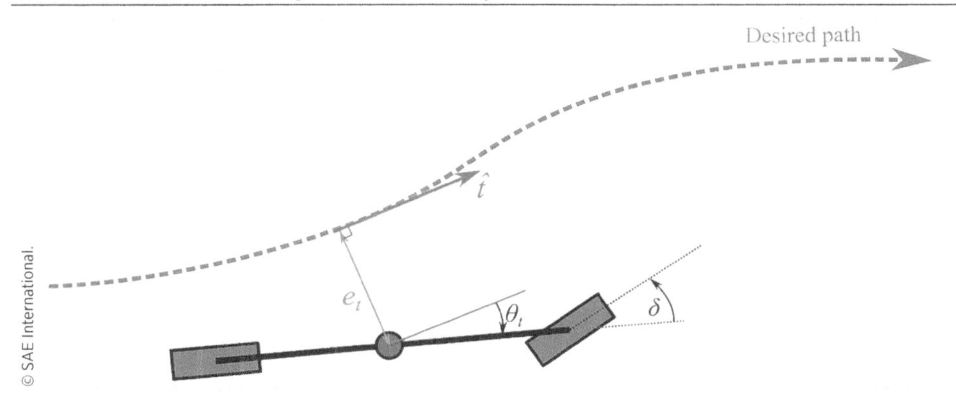

FIGURE 7.18 Trajectory tracking expressed in a vehicle-fixed reference frame, where the x-direction represents the longitudinal direction and the y-direction represents the lateral direction (toward the left of the vehicle). x_e, y_e, and θ_e are the longitudinal, lateral, and heading error between the vehicle's state and the desired trajectory.

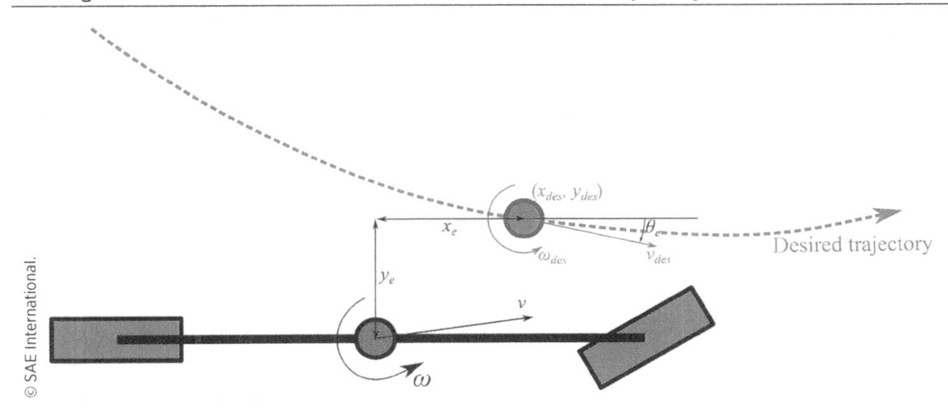

There are limits to the capability of the kinematic trajectory controller. For instance, variations to the target speed v and angular velocity ω may result in uncomfortable or physically impossible changes to steering angle or wheel torque. At higher speeds and accelerations or in emergency maneuvers, it becomes crucial to consider longitudinal and lateral tire slip and the resulting forces. Kinematic vehicle models, which assume that the tires roll without slipping, are insufficient. To properly account for tire forces and vehicle dynamics, a more comprehensive control framework becomes necessary.

Model Predictive Control

Model predictive control (MPC), sometimes called receding horizon control, has become perhaps the most well-established approach to designing trajectory controllers for CAVs (Liu, Lee, Varnhagen, & Tseng, 2017; Ji, Khajepour, Melek, & Huang, 2016). MPC shares similarities with the optimization-based trajectory planners discussed earlier in this chapter. It employs a system model and attempts to find, through numerical optimization, the sequence of control inputs over some forward-looking time horizon that minimizes a cost function generating an optimal path. It then applies the first control input in this optimal sequence. At the next time step, it measures the current vehicle state and resolves the optimization problem over the prediction horizon, and again applies only the control input. Since each step involves solving a constrained optimization problem, MPC is more computationally intensive than other path and trajectory planning methods. However, recent advances in hardware and software capabilities have made MPC use both feasible and practical in real time (Paden, Čáp, Yong, Yershov, & Frazzoli, 2016).

MPC requires a discrete-time model of the system to describe how the vehicle states evolve from time k to time $k + 1$, for all $k \in N$:

$$x(k+1) = F(x(k), u(k)) \tag{5}$$

The system model acts as one of the constraints on the control optimization step of MPC. There are additional constraints on the state $x(k)$ and control $u(k)$:

$$x_k \in X_{free}(k), \quad \forall k \tag{6a}$$

$$p_i(u(k)) < 0, \quad \forall i, k \tag{6b}$$

The first specifies that, at all times in the prediction horizon, the vehicle state must be in the collision-free configuration space X_{free}. The second set of constraints reflects practical limits on the control, such as how fast the steering angle can change, limits to the maximum propulsive or braking torque, etc.

A general form of the MPC cost function J could apply costs g_k to deviations from the reference trajectory and h_k to certain characteristics of the control, such as large or quickly changing steering or longitudinal control inputs. A general form of the optimization problem employed as the control law could then be written as:

$$\min_{u \in U} \quad J = \sum_{k=1}^{N} g_k (x_{des}(k) - x(k)) + \sum_{k=0}^{N} h_k (u(k)) \tag{7a}$$

$$\text{subj. to} \quad x(k+1) = F(x(k), u(k)), \quad \forall k \tag{7b}$$

$$x_k \in X_{free}(k), \quad \forall k$$

$$p_i(u(k)) \leq 0, \quad \forall i, k$$

Sometimes, the last two constraints on the state and control are instead incorporated into the cost function with a high penalty, essentially turning hard constraints into soft ones. This can be helpful if the large number of constraints would otherwise make the problem impractical or infeasible. In the formulation above, g_k, h_k, and p_i may be complicated nonlinear functions, in which case the optimization problem may be computationally expensive or even impossible to solve on board a CAV. Recall that the entire optimization problem must be solved at each time step, though only the first control output $u(k = 1)$ is performed at each step.

Thus, to facilitate real-time implementation, a common approach is to linearize the model, apply quadratic cost functions g and h, linear constraints p_i, and represent U and $X_{free}(k)$ using polyhedra. The result is a quadratic program that can readily and efficiently be solved. The system is linearized around some nominal state x_0 and control u_0, and the linearized model is assumed to provide a reasonable approximation to system dynamics around this point. There are various approaches to linearization in terms of selecting x_0 and u_0, whether to include lateral or longitudinal vehicle dynamics, and deciding how much deviation is allowed before a new linearized model must be computed.

Here a linearized model tracks the deviation of the state from the desired trajectory:

$$z(k) = x_{des}(k) - x(k) \tag{8a}$$

$$z(k+1) = A_k z(k) + B_k u(k) \tag{8b}$$

Employing quadratic costs and linear constraints, the control optimization becomes:

$$\min_{u \in U} \quad J = \sum_{k=1}^{N} z(k)^T G_k z(k) + \sum_{k=0}^{N} u(k)^T H_k u(k) \tag{9a}$$

$$\text{subj. to} \quad z(k+1) = A_k z(k) + B_k u(k) \tag{9b}$$

$$x_k \in X_{free}(k), \quad \forall k$$

$$P_i u(k) \leq 0, \quad \forall i, k$$

Actuation and Actuator Delay

In human-driven vehicles, a driver physically actuates the steering wheel, accelerator pedal, and brake pedal to control the vehicle's trajectory. In the past, there were direct mechanical links between these driver controls and the vehicle's wheels, engine, and hydraulic brakes. Modern vehicles employ electric and electronic systems to augment and sometimes completely replace these mechanical linkages. For instance, electric power steering (EPS) maintains a mechanical steering connection but employs sensors to detect the position and torque of the in-cabin steering wheel and applies an additional assistive torque to help turn the wheels. Full steer-by-wire eliminates the mechanical connection and relies solely on motors to turn the wheels based on the sensed position and torque on the steering wheel. Both types of steering are amenable to automated driving. With EPS, the motion controller's command to apply steering torque would turn both the steered wheels and the in-cabin steering wheel. EPS is essentially a prerequisite to Level 1 automated systems for lateral control such as lane-keeping assist. Full steer-by-wire systems can do away with the steering wheel entirely. Because there is no mechanical connection, steer-by-wire systems can offer a faster response to electronically commanded steering angle changes.

In vehicles with electric propulsion, electronic systems control the motors that generate propulsive wheel torque and facilitate regenerative braking. Even in vehicles with ICEs, electronic throttle control is standard on modern vehicles and is required for basic driver assistance features such as cruise control, and electronically controlled braking is needed for a variety of existing systems such as stability control, anti-lock brakes, and AEB.

As such, the actuators available on modern vehicles are amenable to control by an automated driving agent, though they are not optimized for this purpose. Longitudinal and lateral control actuators can introduce a delay. The steering angle or propulsive torque demanded by the motion controllers cannot be achieved immediately. One way to address this would be to explicitly model and include the dynamics of the individual actuators in the motion control. However, this introduces complexities that would render the control problem impossible to treat in real time. Another way to account for actuator dynamics is to introduce a simplified delay to the controller. The state vector x could be augmented by one or two additional states that reflect a first- or second-order actuator delay. Alternatively, a direct delay could be introduced with some fixed lag time τ_{delay}, such that the actual control at time t is $u_{actual}(t) = u_{commanded}(t - \tau_{delay})$.

End-to-End Automated Driving

Thus far in this chapter, the discussion has focused on the various aspects of a hierarchically organized ADS with distinct behavioral decision-making, trajectory planning, and motion control subtasks. Many approaches employed in the hierarchical framework rely on the recognition of specific features in the environment for behavioral decision-making and trajectory planning. These features, such as traffic control markers, signals, signs, and bounding boxes for other road users, require some manual training and labeling. Furthermore, while HD maps may be used to facilitate driving in many structured environments, they are less useful when there are changes to the environment or its structure, or in many unstructured environments. *End-to-end* learning has been proposed to address some of these challenges by some entities such as Nvidia (Bojarski, et al., 2016). In contrast to the hierarchical description of ADS discussed thus far, an end-to-end learning approach seeks to construct a map directly from the camera and other sensor inputs directly to steering and longitudinal control outputs. Figure 7.19 summarizes the general difference between the two approaches.

FIGURE 7.19 End-to-end automated driving employs a neural network to map sensor inputs directly to vehicle control inputs without explicit hierarchical structure.

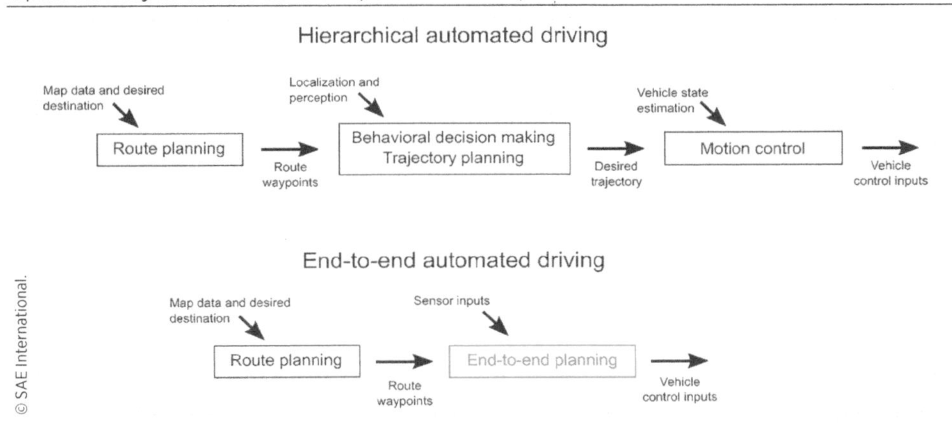

End-to-end automated driving can be considered an extension of the imitation learning-based approaches discussed in section "Learning-Based Methods." However, instead of employing imitation learning only to the behavioral decision-making component of automated driving, end-to-end learning applies to the entire driving task. It learns from human expert demonstrations the appropriate vehicle controls to apply based on what the vehicle sensors see. While conceptually straightforward, end-to-end automated driving is challenging because the extremely variable and high-dimensional nature of the sensor inputs must be mapped to constrained 2D or 3D control inputs. Hierarchical ADS address this by narrowing down the driving task in a sequence of task layers with their own problem formulations. In end-to-end learning, this is entirely addressed by training a CNN (see Chapter 5 for more details on CNNs). An advantage to this approach is that the driving system can be optimized without the need for manual human interpretation of the driving task in a hierarchical manner, which could lead to faster computation and a more efficient software agent. On the other hand, the lack of human interpretation means that validating and verifying the robustness of such a system is more challenging from engineering and regulatory perspectives. Due to these challenges and uncertainty surrounding its behavior in new or unpredictable situations, pure end-to-end automated driving is not being actively pursued. Instead, a more promising approach is to apply imitation learning-based techniques to specific driving subtasks and scenarios.

Summary and Outlook

This chapter presented an overview of various aspects of CAV path planning and motion control, including behavioral decision-making, trajectory planning, and actuation. It organized these tasks in a hierarchical manner. Behavioral decision-making algorithms determine which high-level actions to perform, such as a stop at a stop sign, a lane change, or obstacle avoidance. Trajectory planning involves taking these high-level actions and generating a feasible means to achieve it, ensuring safety while being mindful of factors such as speed and comfort. The motion control task finds the appropriate control inputs and actuation required to track the desired trajectory.

This categorization and separation of tasks is not intended to be descriptive of how all CAV path planning and motion control algorithms operate. In this quickly evolving field, advancements in machine learning and image processing allow developers to learn behaviors and actions directly from human demonstration, as discussed in the section on end-to-end learning. As the technology continues to advance and regulatory bodies begin to take a more active role, it is possible that the various methods currently in the developmental phases will begin to converge around preferred approaches that effectively manage and appropriately balance the multitude of ethical, social, legal, and performance expectations of automated driving.

References

Awad, E., Dsouza, S., Kim, R., Schulz, J., Henrich, J., Shariff, A., ... Rahwan, I. (2018). The Moral Machine Experiment. *Nature*, 563, 59-64.

Bacha, A., Bauman, C., Faruque, R., Fleming, M., Terwelp, C., Reinholtz, C., ... Webster, M. (2008). Odin: Team VictorTango's Entry in the DARPA Urban Challenge. *Journal of Field Robotics*, 25(8), 467-492.

Bast, H., Delling, D., Goldberg, A., Müller-Hannemann, M., Pajor, T., Sanders, P., ... Werneck, R. F. (2016). Route Planning in Transportation Networks. *Algorithm Engineering*, 19-80.

Bojarski, M., Testa, D. D., Dworakowski, D., Firner, B., Flepp, B., Goyal, P., ... Zieba, K. (2016). End to End Learning for Self-Driving Cars. arXiv preprint, arXiv:1604.07316 [cs.CV].

Di Fabio, U., Broy, M., Brüngger, R. J., Eichhorn, U., Grunwald, A., Heckmann, D., ... Nehm, K. (2017). Ethics Commission: Automated and Connected Driving. German Federal Ministry of Transport and Digital Infrastructure.

Galceran, E., Cunningham, A. G., Eustice, R. M., & Olson, E. (2015). Multipolicy Decision-Making for Automated Driving via Changepoint-Based Behavior Prediction. *Robotics: Science and Systems*. Rome, Italy, p. 6.

Hao, K. (2018, October 24). Should a Self-Driving Car Kill the Baby or the Grandma? Depends on Where You're From. MIT Technology Review.

Horizon 2020 Commission Expert Group. (2020). *Ethics of Connected and Automated Vehicles: Recommendations on Road Safety, Privacy, Fairness, Explainability and Responsibility*. Publication Office of the European Union, Luxembourg.

Ji, J., Khajepour, A., Melek, W. W., & Huang, Y. (2016). Path Planning and Tracking for Vehicle Collision Avoidance Based on Model Predictive Control with Multiconstraints. *IEEE Transactions on Vehicular Technology*, 66(2), 952-964.

Kammel, S., Ziegler, J., Pitzer, B., Werling, M., Gindele, T., Jagzent, D., ... Stiller, C. (2008). Team AnnieWAY's Automated System for the 2007 DARPA Urban Challenge. *Journal of Field Robotics*, 25(9), 615-639.

Kolski, S., Ferguson, D., Bellino, M., & Siegwart, R. (2006). Automated Driving in Structured and Unstructured Environments. *Proceedings of IEEE Intelligent Vehicles Symposium*. Meguro-Ku, pp. 558-563.

Lima, P. F. (2018). Optimization-Based Motion Planning and Model Predictive Control for Automated Driving: With Experimental Evaluation on a Heavy-Duty Construction Truck. Doctoral dissertation, KTH Royal Institute of Technology.

Liu, C., Lee, S., Varnhagen, S., & Tseng, H. E. (2017). Path Planning for Automated Vehicles Using Model Predictive Control. *IEEE Intelligent Vehicles Symposium*. Los Angeles.

Paden, B., Čáp, M., Yong, S. Z., Yershov, D., & Frazzoli, E. (2016). A Survey of Motion Planning and Control Techniques for Self-Driving Urban Vehicles. *IEEE Transactions on Intelligent Vehicles*, 1(1), 33-55.

Sadigh, D., Landolfi, N., Sastry, S. S., Seshia, S. A., & Dragan, A. D. (2018). Planning for Cars that Coordinate with People: Leveraging Effects on Human Actions for Planning and Active Information Gathering over Human Internal State. *Automated Robots*, 42(7), 1405–1426.

Schwarting, W., Alonso-Mora, J., & Rus, D. (2018). Planning and Decision-Making for Automated Vehicles. *Annual Review of Control, Robotics, and Automated Systems*, 1, 187-210.

Thornton, S. M. (2018). Automated Vehicle Motion Planning with Ethical Considerations. Doctoral dissertation, Stanford University.

Urmson, C., Anhalt, J., Bagnell, D., Baker, C., Bittner, R., Clark, M. N., … Ferguson, J. (2008). Automated Driving in Urban Environments: Boss and the Urban Challenge. *Journal of Field Robotics*, 25(8), 425-466.

VOL
RADIO
MEDIA
MAP
NAV
FRONT
REAR
PASSENGER AIR BAG
A/C
CLIMATE
MODE
DUAL
CleanAir
TEMP
AUTO
TEMP

8

Verification and Validation

The issue of safety of CAVs deployed on public roads, referred to as "operational safety,"[1] is perhaps the main driver of enthusiasm for CAVs by various stakeholders, including the automotive industry, government, safety advocates, academia, and the general public. CAVs have the potential to improve operational safety, but this potential must be proven. There are two steps to determining the operational safety of CAVs:

1. Determining how the level of operational safety will be measured
2. Establishing what level of operational safety is acceptable

The second step is outside the scope of this chapter, as there is currently no consensus on what is "safe enough" for CAVs, and this will involve a variety of stakeholders: government, the general public, industry, ethics experts, etc. In other words, this is not an engineering initiative, but rather a multi-disciplinary endeavor.

The first step is the subject of this chapter and a crucial area for making sure that CAVs fulfill the operational safety promise. The need for such determination is common across the automotive industry for conventional vehicles, and indeed for all consumer products, and is commonly known as verification and validation (V&V). A robust V&V process that can provide assurance to the public that CAVs are sufficiently safe for commercialization is essential, and the current status of this process is the subject of this chapter. It should be noted that CAVs will generally have to meet Federal Motor Vehicle Safety Standards (FMVSS) like all conventional vehicles in the United States, although there may

[1] Operational safety refers to when the vehicle is in operation on public roads. Other aspects of safety such as incident safety, which refers to what happens when the vehicle is in a collision, i.e., airbag deployment, are also important but outside of the scope of this chapter.

be exceptions in cases like a Nuro delivery vehicle where no steering wheel or passenger cabin is inherent in the vehicle design.

Definition

It is first important to be clear what, exactly, is meant by V&V. V&V, depicted as part of a design process in Figure 8.1, are related but separate tests of a system:

- Verification tests whether the system meets the specifications (error-free, engineered well, etc.) This is an objective process that includes hard metrics and proven design analysis tools.
- Validation tests whether the specifications meet the application needs. This is a much more subjective process that includes softer metrics like user evaluation.

FIGURE 8.1 V&V depiction.

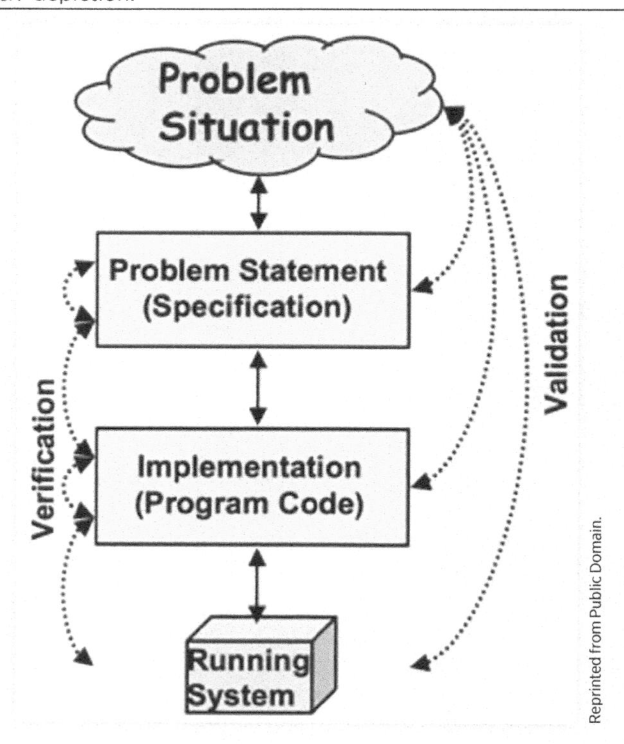

The V&V process can also be thought of as part of the well-established V-Model of product development, as shown in Figure 8.2.

FIGURE 8.2 V-Model of product development.

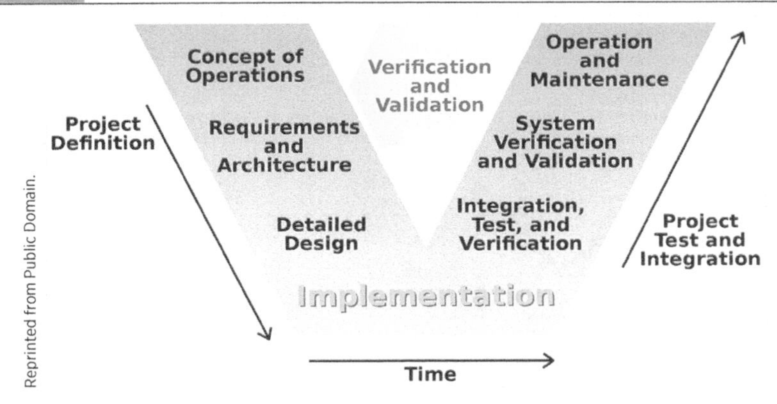

Reprinted from Public Domain.

The process starts on the left-hand side of Figure 8.2, with Project Definition, which includes developing:

1. Concept of Operations (ConOps) that describes the characteristics and usage of the product
2. Requirements and Architecture that set the expectations and basic design of the product
3. Detailed design of the product

With the design complete (note that the process can be iterative so this step is often completed several times), the Project Test and Integration steps are then conducted, including:

1. Integration, Test, and Verification of the sub-systems
2. System V&V to ensure that all requirements are met and the design is acceptable
3. Operation and Maintenance in which the product has been commercialized or deployed and its operation must be maintained

A more detailed version of the V-Model is shown in Figure 8.3. This modified version shows explicitly the roles of V&V and at what point in the development process each occurs (including explicitly specifying traceability), along with sub-system and component distinction from the overall system.

FIGURE 8.3 More detailed version of the V-Model.

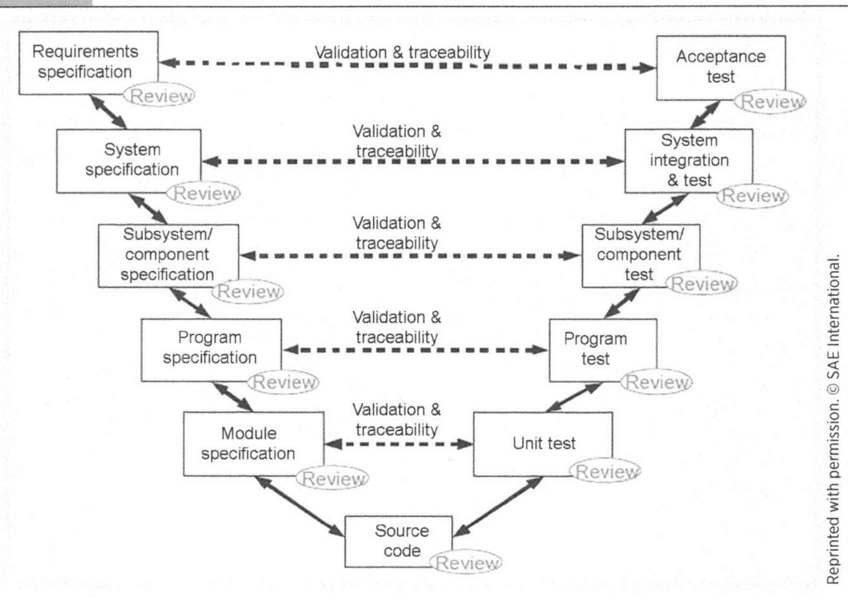

The left-hand side of the "V" can be thought of as the design and development stage while the right-hand side is the test and development stage. There are many available methods that can be used for both stages, as discussed below. There are advantages associated with each method, and all are used in the V&V of CAVs but also of conventional vehicles as well. The methods should not be seen as alternatives, but rather as complementary methods that are used in aggregate to establish the level of safety of the CAV. This will be discussed further in section "Safety Case."

Design and Development Methods

Design and Development methods make use of accepted design techniques to achieve a product that meets requirements. These methods will be described briefly in this section but are not the focus of this chapter. The referenced documents can be explored for further information.

Systematic Process: A product development process that follows best practices in how the product is defined, developed, verified and validated, deployed, and maintained [quoted from (IEEE, 2021)]. Example best practice documents include ISO 26262 [International Organization for Standardization (ISO), 2018] and ISO/PAS 21448 [International Organization for Standardization (ISO), 2019]. These two documents on functional safety and on safety of the intended functionality, respectively, will be discussed in more detail later in the chapter.

Safety-by-Design-Architectures: A product design method showing compliance with accepted reference architectures [quoted from (IEEE, 2021)]. Example best practice documents include SAE J3131 (SAE International, In preparation) and ISO/TR 4804 [International Standards Organization (ISO), 2020].

Formal Methods: A product design method that uses mathematically rigorous techniques to model complex systems, particularly safety-critical systems, in a precise and unambiguous manner [quoted from (IEEE, 2021)].

Robustness Analysis: A product design technique that analyzes system sensitivity to variability, including variable environmental conditions, number and type of traffic participants, sensor noise, etc. [adapted from (IEEE, 2021)].

Test and Validation Methods

Test and Validation methods can be used once a design has been completed (even if only partially and/or as a prototype) to measure the performance of the design. There are several different test and validation method types:

1. Destructive testing to measure the design's robustness to stress.
2. Nondestructive testing to evaluate the design without causing damage. Performance testing can be considered to be a type of nondestructive testing.
3. Fault injection testing to measure system behavior to forced faults in the software or hardware of the design.

Test and Validation methods can also be separated into two categories:

1. Virtual versus real
2. Component versus system

The first category describes whether the testing is simulated or done with actual equipment, which can be considered as a continuum between fully simulated and fully real known as "X-in-the-Loop" (XiL; more on the different "X" types below), as shown in Figure 8.4. Virtual testing has increased considerably as computational power has improved and virtual tools have progressed. Since virtual testing is much cheaper and faster than real testing, many CAV developers are conducting the vast majority of testing in simulation, followed by XiL testing, and last by fully real testing (Tahir & Alexander, 2020). This is no surprise and the trend is likely to continue as simulation tools continue to be developed. However, any simulated testing must be validated, and fully real testing allows for a fidelity of 1 and no need for validation.

FIGURE 8.4 XiL testing methods compared against test drives.

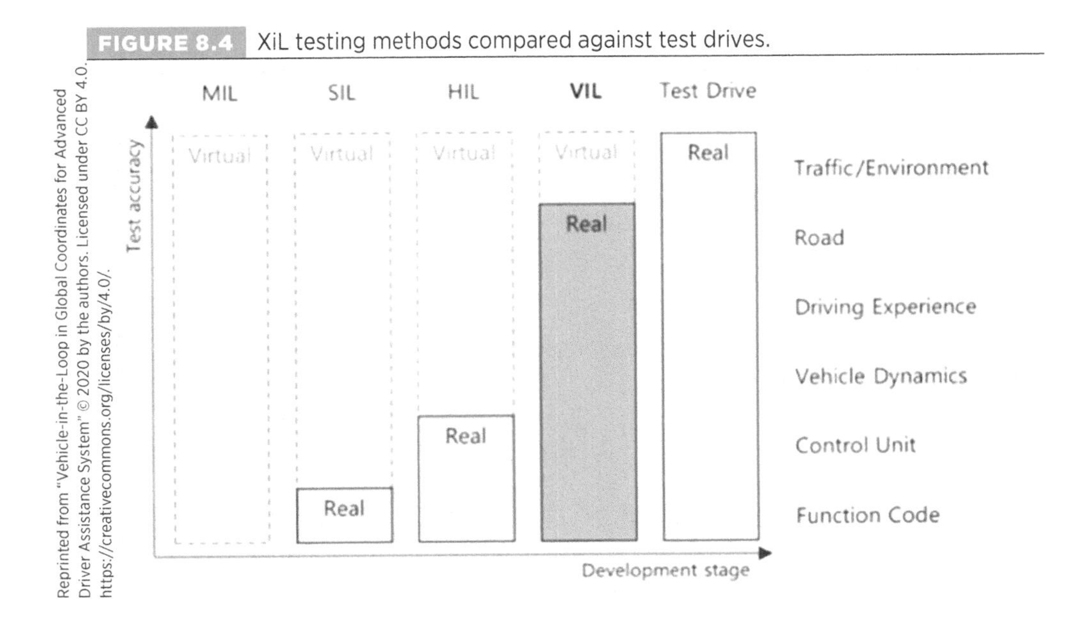

The second category describes whether the entire system is being tested (i.e., the CAV) or if one of its constituent components is being tested. Both types will be described in this section, but the system-level testing will be the focus of this chapter and described in greater detail. The referenced documents can be explored for further information.

Model-in-the-Loop (MiL) Testing: A method of testing in which a plant model or control system is tested in a modeling environment to perform V&V of a model. An example document is (Plummer, 2006).

Software-in-the-Loop (SiL) Testing: A method of testing in which executable code is tested in a modeling environment to perform V&V of software. An example document is (Ahamed, Tewolde, & Kwon, 2018).

Simulation Testing: A product testing method that uses a simulation platform to place the CAV model in various scenarios to test the operational safety performance. Example best practice documents include ASAM OpenDrive (ASAM, 2021) and (Do, Rouhani, & Miranda-Moreno, 2019). Simulation testing could be considered as comprising MiL and SiL.

Hardware-in-the-Loop (HiL) Testing: A test methodology where an electronic control unit (ECU) is connected to a plant model in a closed loop in order to evaluate the response of the controller, typically used later in the development process and performed in real time in order to test the functions of the ECU but before testing in the final system [quoted from (SAE International, In preparation)]. Example documents include (Ma, Zhou, Huang, & James, 2018) and (Xu, et al., 2017).

Driver-in-the-Loop (DiL) Testing: Execution of the target software on prototypical or target hardware in the target CAV or a mockup, and the environment is modified with virtual stimuli, whereas the driver's reaction influences the vehicle's behavior [adapted from (International Standards Organization (ISO), 2020]. When the driver is an ADS, this type of testing is also known as Vehicle-in-the-Loop (ViL) testing. An example document is (Tettamanti, Szalai, Vass, & Tihanyi, 2018).

Closed Course Testing: A product testing method that involves the operation of a CAV at a closed course test facility for the purpose of testing and validating its behavior in controlled scenarios by comparing the outcome to an expected outcome (e.g., a collision vs. maintenance of a safety envelope around the subject vehicle) [adapted from (IEEE, 2021)]. An example document is (Omidvar, Pourmehrab, & Emami, 2018).

Public Road Testing: A product testing method that involves operation of the CAV on public roads for the purpose of testing and validating its behavior in real-world (i.e., non-controlled) scenarios by comparing the outcome to an expected outcome (e.g., a collision vs. maintenance of a safety envelope around the ego vehicle) [adapted from (IEEE, 2021)]. Example documents include ECE/TRAN/WP.29 (United Nations Economic Commission for Europe, 2020), SAE J3018 (SAE International, 2020), and [Automated Vehicle Safety Consortium (AVSC), 2020].

Challenges

The V&V process of a product can vary widely depending on the particular product and also a number of factors, including:

- Regulatory/standard stringency
- Technology readiness
- Consumer acceptance
- Developer risk acceptance

In the case of ADS, there are no regulations or standards (currently) that dictate how an ADS development and/or subsequent V&V process must be conducted, which type or amount of testing must be conducted, and what level of performance is required for the ADS. The U.S. federal regulatory body, the NHTSA, released an Advance Notice of Proposed Rulemaking in November 2020 (USDOT - NHTSA, 2020) as a first step toward developing the V&V process, but there is still far to go before there is full regulation. In Europe, the United Nations Economic Commission for Europe (UNECE) released the Automated Lane Keeping Systems regulation for low-speed features in limited ODDs (United Nations Economic Commission for Europe, 2020). The European continent's CAV regulatory environment is ahead of that in the United States, but a comprehensive set of regulations is still outstanding.

In addition to regulations and standards, OEMs normally draw upon their knowledge of best practices and their own experience to develop their products to assess safety and quantify risk. There are few common best practices for CAV V&V and with little outside guidance, CAV developers are devising their own, proprietary processes. One example of how approaches will vary is that there is no agreed-upon set of scenarios for any given ODD, and there is no clear path to achieving such a set that is representative of the reasonably foreseeable scenarios that a CAV could encounter in said ODD. Even if an agreed-upon set could be achieved, having predefined tests have proven to be a problem already in V&V and compliance testing of conventional vehicles—see, for instance, the VW diesel scandal—and this is with "dumb" vehicles. The problem of "gaming" the tests could grow when the vehicle employs AI and can learn how best to pass a predefined test.

Without a common process with publicly available data, it is difficult to know the level of safety that is actually being achieved. Part of the V&V process development must include agreement on the type and amount of data to be shared with a regulator, but also with the public. Another part (discussed in section "Evaluation Methods") is deciding the minimum level of safety that must be achieved and whether all CAVs must navigate a given scenario in the same manner or if the evaluation will be restricted to safety only. It is clear that CAV stakeholders (government, industry [including standards bodies], academia, and the public) have a lot of work to do to develop the best practices-standards-regulation pipeline that is normally used to develop a V&V process.

Technology readiness is another technical challenge for CAVs. The sensors described in Chapter 4, AI in Chapter 5, sensor fusion algorithms in Chapter 6, and path planning techniques in Chapter 7 have improved significantly, but performance and cost objectives have not been entirely met. With such flux in performance, developing a V&V process is difficult, although the widespread objective is to have a process that is "technology neutral" and performance based. This neutrality is important because various approaches to CAV design have been undertaken by the CAV developers to date, such as the push by some developers to use camera-only sensor suites and not RADAR and/or LIDAR like most others (Schmidt, 2021).

The V&V process development is further complicated by the existence of the different SAE levels of automation from SAE J3016. There will have to be different processes for each level due to the unique characteristics. For example, SAE Level 2 and Level 3 each pose unique challenges because of the human-machine interactions. Some of the outstanding questions are:

- How can the human be deemed to be sufficiently "situationally aware" in order to take over the DDT?
- What is the minimum time required for the driver takeover and how to measure the "quality" of the takeover?

It should be noted that the V&V of CAVs will be performed for both performance and safety (although the rest of this chapter will focus on the safety aspect). The performance objective relates to the CAV's ability to perform the DDT within the ODD (i.e., behavioral competency). The safety objective relates to the CAV's ability to perform the DDT within the ODD in a manner that is safe for the CAV and other entities in the environment. These objectives aren't necessarily complementary. For example, an aggressive-driving CAV could cause accidents while being able to successfully navigate from A to B. The combined safety and performance objectives of V&V are to ensure that the CAVs can safely execute a set of behavioral competencies for scenarios that are seen by vehicles on public roads. However, the set of behavioral competencies for any particular level of automation, ODD, and fallback method has yet to be finalized. The California Partners for Advanced Transportation Technology (PATH) program at UC Berkeley (founded in 1986) developed 28 minimum behavioral competencies for AVs that were adopted by the NHTSA for Levels 3-5 (University of California PATH Program, 2016). Waymo developed an additional 19 behavioral competencies in 2017 in the submitted VSSA to the NHTSA (Waymo, 2017). The full set of behavioral competencies could be much larger than these initial proposals and will have to be agreed upon for the V&V process.

A V&V process will have to be sufficiently robust and transparent to instill public confidence that public safety is being maintained with the deployment of CAVs on public roads. Some companies have perhaps exaggerated the abilities of their vehicles with press releases on contemporary timeframes for CAV deployment (Hyperdrive, 2020). The result of the V&V process is a vehicle with automation feature(s) that can be sold to a consumer. This requires an unprecedented amount of trust to be placed in automation (i.e., the vehicle's ADS) to safely handle the DDT, but will also require unprecedented user knowledge and understanding for SAE levels below 4/5 where the human driver remains in the loop.

Public acceptance becomes complicated if there are deviations from the normal V&V progression. Since V&V of a product is usually done once, if a product redesign occurs, the V&V process normally has to be conducted again. For CAVs with OTA capabilities, determining when the CAV has to be re-tested, i.e., the point at which the results of the original process no longer apply, will be related to the requirement to maintain public confidence. It should also be noted that OTA updates have not always been popular with Tesla owners, with Tesla pioneering this practice in the automotive industry (Siddiqui, 2021). The logistics of re-doing the V&V process for CAVs already in the hands of customers will need to be developed and agreed upon by multiple parties, including the CAV seller, regulatory body, and CAV customer.

Public confidence in CAV safety will strongly depend on the number of accidents involving CAVs as they are being tested on public roads before deployment. However, public road testing (discussed in more detail in section "Public Road Testing") is widely seen as a likely crucial aspect of any V&V process. How this public road testing is conducted will have a massive impact on public acceptance of CAVs.

Related to the public acceptance aspect of CAV V&V is CAV developer risk tolerance and how public road testing is both a necessary component of the V&V process and also poses a significant liability. Incidents such as the death of a pedestrian in Tempe, AZ in 2018 [National Traffic Safety Board (NTSB), 2019] by a prototype Uber CAV caused a major change to operations in Uber's CAV division (which was ultimately sold to Aurora in late 2020). CAV developers must balance the need to validate and accumulate real-world miles and the benefits of testing on public roads while at the same time being aware of the risk associated with this testing. Without a set of testing standards or scenarios, it is impossible to assess or price risk or predict actual CAV performance in the real world. This leads to questions of how the insurance industry will participate, and CAV developers have largely been self-insuring while CAVS are being tested on public roads. The appetite for this risk could change if there are more incidents that change public opinion, regulation is introduced, and/or if any lawsuits result in large monetary judgments against CAV developers.

Overall, V&V of "intelligent" systems like a CAV is inherently complex and fundamentally different compared to passive safety features and (relatively) simple active safety ones. The former generally have accepted V&V processes while the latter's processes are still largely in development. The V&V process for CAVs will likely be a multi-year, multi-jurisdictional progression. Chapter 9 contains more discussion of the regulatory environment and the outlook for a required V&V process.

Test and Validation Methodology

The Test and Validation methodology (that uses the methods described earlier in the chapter) will be discussed in further detail and will be the subject of the rest of this chapter. The four (4) questions that must be answered for the Test and Validation methodology for which operational safety is the objective are:

1. What is being measured when testing, i.e., what are the operational safety metrics?
2. How is testing being conducted, i.e., which of the methods from section "Test and Validation Methods" is being used?
3. How are the test results interpreted/evaluated?
4. What are the pass/fail criteria?

The four elements of the Test and Validation methodology will be discussed in further detail in the following sections.

Operational Safety Metrics

There is an age-old adage that states "Measure before you manage." Determining what, exactly, is to be measured is critical. There have been operational safety assessment (OSA) metrics used for decades for evaluating the performance of human-driven vehicles, and the OSA metrics can generally be classified as either leading or lagging. Lagging metrics are measurements of actualized OSA outcomes while leading metrics are measurements that can (depending on the strength of data to support the contention) predict the OSA outcome (Automated Vehicle Safety Consortium, 2021).

Common lagging metrics are collision incidents and rules-of-the-road violations. These are actualized outcomes of a vehicle (whether driven by a human or ADS) navigating a given scenario. These metrics are useful in that there is little ambiguity about their meaning: if there is a collision, this is certainly an example of an unsafe situation. The situation is a bit more complicated for rules-of-the-road violation, where, in certain circumstances, e.g., when being directed by a police officer to drive on the road shoulder to avoid an accident scene, a violation is not an indication of an unsafe situation. This latter state of affairs is rare, however.

A large number of leading metrics have been proposed by various research groups, mostly before CAVs were considered. A recent paper (Mahmud, Ferreira, Hoque, & Tavassoli, 2017) summarized leading OSA metrics according to different bases (temporal, distance, and deceleration):

- Temporal-based metrics: TTC, Extended Time-to-Collision (ETTC), Time-Exposed Time-to-Collision (TET), Time-Integrated Time-to-Collision (TIT), Modified TTC (MTTC), Crash Index (CI), Time-to-Accident (TA), Time Headway (THW), and Post-Encroachment Time (PET).
- Distance-based metrics: Potential Index for Collision with Urgent Deceleration (PICUD), Proportion of stopping Distance (PSD), Margin to Collision (MTC), Difference of Space Distance and Stopping Distance (DSS), Time-Integrated DSS (TIDSS), and Unsafe Density (UD).
- Deceleration-based metrics: Deceleration Rate to Avoid a Crash (DRAC), Crash Potential Index (CPI), and Criticality Index Function (CIF).

More recently, leading safety envelope-based OSA metrics (defined later in the chapter) with CAVs in mind have been proposed by Intel/Mobileye [based on its Responsibility-Sensitive Safety (RSS) framework], Nvidia with its Safety Force Field (SFF), and researchers at the University of Michigan Transportation Research Institute (UMTRI) with the Model Predictive Instantaneous Safety Metric (MPrISM).

While many OSA metrics have been proposed, no consensus had been reached on what OSA metrics are necessary. Researchers at the IAM (the authors of this book are members who are leading this research) proposed a comprehensive set of OSA metrics following an extensive literature review (Wishart et al., 2020). The objective was to develop a foundational set of OSA metrics for both human-driven vehicles and CAVs that includes existing, adapted, and novel metrics. In a follow-up paper, the IAM researchers proposed a taxonomy for OSA metrics and used simulation studies to evaluate the safety envelope-based metrics. The SAE V&V Task Force has further developed a proposed taxonomy for a Recommended Practice on OS metrics under development (SAE J3237, In preparation). The proposed taxonomy is shown below in Figure 8.5.

FIGURE 8.5 OS metrics taxonomy proposed by SAE (SAE J3208, In preparation).

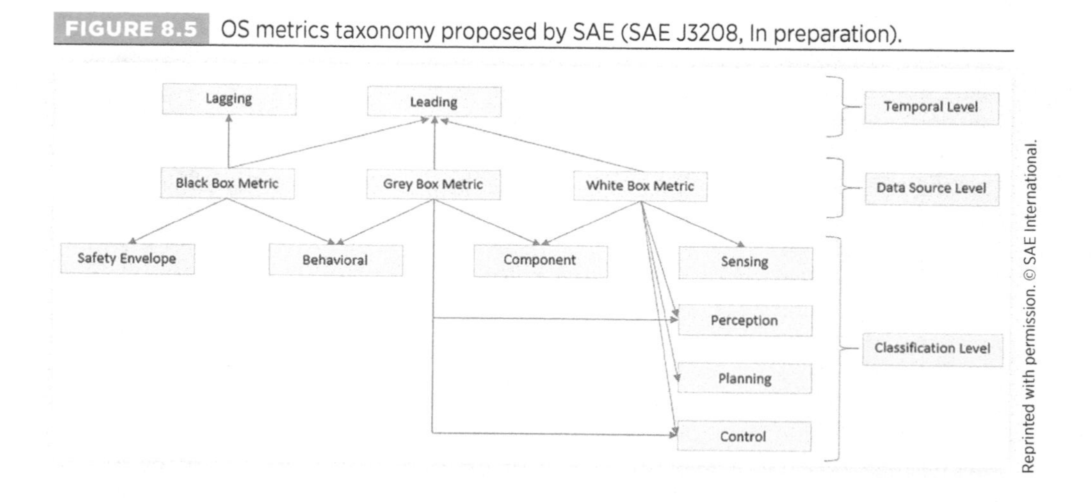

The top level of the taxonomy is the Data Source Level, and it indicates where the data for the measurements originate. The premise for this taxonomy is that depending on the use case (i.e., if it is a CAV developer using metrics, a third-party evaluator, or a regulatory body), ADS data may be more or less accessible:

- Black Box Metric: An OSA measure that does not require any access to ADS data. This could be from an on-board or off-board source. ADS data may enhance the accuracy and precision of the measurement(s). An example Black Box metric would be a collision incident.
- Gray Box Metric: An OSA measure that requires limited access to ADS data. An example Gray Box metric would be one that indicates whether the ADS is executing the DDT or not.
- White Box Metric: An OSA measure that requires significant access to ADS data. An example White Box metric would be one that measures the precision of the sensor fusion algorithm output.

Gray Box metrics are an attempt at a compromise of useful data that do not require proprietary ADS data to be shared.

The second level of the taxonomy is the Classification Level, which differentiates between the various types of metrics:

- Safety Envelope Metric: An OSA measure of the CAV's maintenance of a safe boundary around itself. This includes situations that may not be within the subject vehicle's control. An example Safety Envelope metric is the Minimum Safe Distance Violation (MSDV) proposed by the IAM and based on the RSS framework developed by Intel/Mobileye (Wishart J., et al., 2020).
- Behavioral Metric: An OSA measure of an improper behavior of the subject vehicle. An example Behavioral metric would be one that measures the aggressiveness of the CAV, such as hard accelerations/decelerations.
- Component Metric: An OSA measure of the proper function of CAV components. An example Component metric would be one that indicates compliance of the event data recorder (EDR) to a regulation.
- Sensing Metric: An OSA measure of the quality of data collected by the CAV sensors about the CAV environment. An example Sensing metric would be one that measures the camera resolution of the CAV's cameras.
- Perception Metric: An OSA measure of the quality of interpretation of data collected by the CAV sensors about the CAV environment. An example Perception metric would be one that measures the CAV's ability to perceive objects within its environment.
- Planning Metric: An OSA measure of the ability of the CAV to devise a suitable trajectory through the CAV environment. An example Planning metric would be one that measures the CAV's planned trajectory for object avoidance.
- Control Metric: An OSA measure of the ability of the CAV to execute the planned route devised by the CAV. An example Control metric would be a measure of the actuation of the throttle to match the acceleration corresponding to the planned trajectory.

The third level of the taxonomy is the Temporal Level, which differentiates between metrics that are measured before and after an operational safety outcome has occurred:

- Leading Metric: An OSA measure that predicts an operational safety outcome. An example Leading metric would be one that measures the CAV's adherence to a safety envelope, i.e., maintaining minimum distances from safety-relevant entities in its environment.

- Lagging Metric: An OSA measure that quantifies an operational safety outcome or adherence to traffic laws. An example Lagging metric would be one that quantifies a collision in which the CAV is involved.

Using this taxonomy, the IAM's proposed list of OSA metrics is shown in Table 8.1. It should be noted that there are only Black Box and Gray Box metrics; this was done purposefully for the use cases where little to no ADS data will be available. The proposed taxonomy is also shown with the Classification level explicitly; the Temporal level is shown with Lagging metrics in italics. The details of each OSA metric will not be explored here. Interested readers can see (Elli, Wishart, Como, Dhakshinamoorthy, & Weast, 2021). It should also be noted that the set of metrics proposed here is not standardized and conversations surrounding the operational safety metrics to be used in CAV development are in early stages; as such, the safety metrics used to design CAVs are subject to the rapidly evolving industry.

TABLE 8.1 IAM proposed OSA metrics set.

Black box metrics		**Gray box metrics**	
Safety envelope	**Behavioral**	**Behavioral**	**Planning**
Minimum Safe Distance Violation	*Collision Incident*	*Human Traffic Control Violation Rate*	*Achieved Behavioral Competency*
Proper Response Action	*Rules-of-the-Road Violation*	**Perception**	**Control**
Minimum Safe Distance Factor	Aggressive Driving	*Human Traffic Control Perception Error Rate*	*ADS DDT Execution*
Time-to-Collision Violation		Minimum Safe Distance Calculation Error	
Modified Time-to-Collision Violation			
Post-encroachment Time Violation			
Time Headway Violation			

The Advanced Vehicle Consortium (AVSC) is an SAE-based consortium with traditional automotive OEMs but also ADS developers and rideshare companies. The AVSC was formed with the purpose of developing thought leadership by CAV industry companies. Among several other white papers, the AVSC released an OSA metrics proposal that was quite similar in philosophy to that of the IAM (Automated Vehicle Safety Consortium, 2021). With two best practice documents exhibiting such agreement, the needed consensus on which OSA metrics should be included has begun to form. An SAE Recommended Practice document (one level below a Standard for SAE) (SAE J3237, In preparation) is currently being developed on OSA metrics that reflects the consensus between the work of the IAM and AVSC by the SAE V&V Task Force under the ORAD Committee [the V&V Task Force is chaired by one of the authors (Wishart)]. The OSA metrics list of the Recommended Practice document has yet to be finalized, but the Task Force has agreed on desirable characteristics of an OSA metric:

- OSA metric definition is transparent (i.e., not proprietary)
- Useful throughout the ODD(s)
- Minimize subjectivity
- Defines a good balance between safety and usefulness by minimizing false negatives (i.e., OSA metric indicates a safe situation when it's actually unsafe) and false positives (i.e., OSA metric indicates an unsafe situation when it's actually safe)
- Needs to be implementable for a particular use case (i.e., easily measurable, does not require intensive compute, does not require proprietary IP or data)
- Is comprehensive of the kinematics components of the DDT
- Is dynamic (i.e., adjusts as the situation changes)

Note that the last two characteristics are applicable to the Safety Envelope metrics only. The objective will be to arrive at a final list of OSA metrics that can then be used to gather data from the test methods of the following section (section "Test Methods") and then these data can be inputs to the evaluation methodology of section "Evaluation Methods".

Test Methods

The various types of test methods were introduced in section "Test and Validation Methods," but the most important system-level methods will be discussed further in this section. There is widespread agreement that a CAV V&V process will include some combination of scenario-based simulation, closed course, and public road testing. Simulation is likely to be the main area of focus (due to efficiency and cost), but testing is likely to be iterative, for example, scenarios experienced in closed course testing and public road testing will be inputs to the simulation testing to further examine the scenario and improve the ADS response. This will be discussed further in section "Safety Case."

Simulation Testing

Simulation testing is the virtual replication of physical objects and phenomena. The "ego" view of a CAV in simulation is shown in Figure 8.6. Simulation testing is the main V&V test and validation method in terms of percentage of overall testing, although it is not as visible as public road testing or even closed course testing. CAV developers can test orders of magnitude more scenarios in simulation than on public roads or closed courses. For example, as of April 2020, Waymo had tested their CAVs for 20 million miles on public roads while testing their models for 15 billion miles through simulation. In fact, Waymo tests 20 million miles per day in simulation, which was the same as the total amount on public roads at that point in 2020 (Waymo, 2020). The daily rate has likely increased since April 2020, as the daily rate was 8 million miles per day in 2017 (Waymo, 2017).

FIGURE 8.6 Simulated ego view of a CAV in traffic.

The CAV V&V process will require large amounts of annotated data in a wide variety of scenarios for training the AI algorithms. Variations in environmental conditions such as weather, lighting, and road conditions must be included. Additionally, variable CAV designs mean that different sensor data must be collected and annotated. Gathering all of these data through physical testing is expensive and time consuming, and likely to require multiple CAVs, which presents operations and maintenance challenges. In contrast, a high-fidelity simulation tool can augment and improve AI algorithm training to make development and subsequent V&V testing more efficient and do so for multiple CAV developers. The large amount of data collected during simulation testing can thus all be useful for V&V purposes.

Simulation is highly beneficial in allowing scalable testing of as many scenarios as the testers can devise, even if the scenario has never been seen in the real world. Additional traffic participants are easily added, and the scenario conditions from CAV subject vehicle speed to time of day can be controlled. The scenarios in simulation testing are often sourced from the most challenging scenarios encountered via public road driving with a prototype CAV with the ADS completing the DDT (known as safety driving if a human safety driver is present to monitor the ADS operation) or with the sensors of the prototype CAV collecting data, and the perception and planning sub-systems running and providing output but with a human driver in control (known as shadow driving). Simulation testing also allows for simple and rapid ADS robustness evaluation by modifying various parameters of a scenario and evaluating whether small changes in parameters result in step changes in ADS response.

There are many approaches to developing a simulation tool, with some organizations developing open-source tools, while others develop commercial or proprietary applications. A comprehensive list of the major simulation tools is included in (Wishart J., et al., 2020). Some characteristics of a useful simulation tool include (Fadaie, 2019):

1. An environment based on an HD map (such as those described in Chapter 2) for accuracy and usefulness when paired with closed course testing or public road driving.

2. Flexible, diverse, and accurate models of the elements of the scenario, such as vehicle models, including human driver models for human-driven vehicles and all the components of an ADS for CAVs, as well as map semantic models such as signage and traffic control equipment, weather conditions and lighting, and other objects in the environment. A view of a scene in four weather conditions in the CARLA simulator is shown in Figure 8.7. Real-world imperfections such as sensor noise also increase the realism of a simulation tool.
3. The simulation is deterministic such that if an initial state is specified, the end result will always be the same.
4. The tool can be extended via crowdsourcing or upgrades for constant improvement.
5. It is easy to parameterize a scenario and run many variations with slight changes to one or more parameters.
6. The results of any scenario output are clearly observable so that evaluations can be made.
7. The tool is easy to use and control, is intuitive, and has compatible interfaces between components, both internal modules and external modules.
8. The tool is capable of high throughput of scenario results (although this will be inherently related to complexity, fidelity, and scope of the simulation).
9. Real-world experience replay is available so that actual behaviors can be visualized and examined more thoroughly and especially so that improvements can be clearly elucidated.

FIGURE 8.7 Depiction of a scene in four different environmental conditions in CARLA.

The main limitation of simulation is fidelity (to reality). Progress has been made in the modeling of elements such as environmental conditions, sensors, human drivers, and vehicle dynamics. Fidelity should not be confused with "photo-realistic," with the latter required only when perception algorithms are being tested; if the sensor data are provided, the simulation need not be photo-realistic. Likewise, if the path planning algorithm is the focus of the test, the fidelity of behaviors and dynamics of the other traffic participants is crucial (Fadaie, 2019). The trade-off with higher fidelity is that the computational power requirement to run a scenario is higher, so more expensive equipment is required and/or fewer scenarios can be run over the same amount of time. Real-world validation will be always required to ensure that simulated results are representative; however, as the simulation tools and their constituent component HD maps and models become higher fidelity, validation may be simplified and overall trust in the simulation tools will increase. A key aspect of the section "Safety Case" is understanding just how much validation is required of the simulation results.

Closed Course Testing

Closed course testing involves a facility that allows CAV testing to occur in a safe, controlled, and non-public environment, often with specialized equipment and/or a traffic environment that is meant to mimic the public road environment. A closed course in Michigan known as Mcity designed specifically for CAV testing is shown in Figure 8.8. Closed course testing is often used to validate the results from simulation testing and to ensure that the ADS can achieve sufficient behavioral competency to be deployed for public road testing.

FIGURE 8.8 Mcity closed course testing facility.

Photo by Eric Bronson, Michigan Photography, University of Michigan.

Closed course testing is the intermediate step between simulation and public road testing in several ways. The fidelity is intermediate because the equipment such as the pedestrian surrogate shown in Figure 8.9 is designed to mimic a human pedestrian but does not succeed entirely. The controllability is intermediate because scenarios can be designed and executed, but the capacity to conduct any scenario that could be imagined is limited to time, cost, and physics constraints. Closed course testing is also often the intermediate step chronologically as this testing is done to refine the ADS after simulation but before deployment on public roads.

FIGURE 8.9 Closed course testing with a pedestrian surrogate.

Closed courses allow for whatever scenarios testers can devise, within limits of the facility and the available equipment. Conventional proving grounds owned by OEMs (often individual OEMs in the United States and shared by two or more OEMs in Europe) can provide basic facilities for CAV testing. However, facilities like Mcity or the American Center for Mobility (ACM, also in Michigan) that are specifically designed for CAV testing purposes with simulated buildings, intersections, traffic control, V2X communications for connectivity testing, etc., will obviously have greater capabilities for CAV-related testing. However, no single facility is likely to have the infrastructure or equipment to be capable of providing all scenarios. Further, the environment itself can be difficult to provide: facilities like the Exponent Test and Engineering Center (at which three of the authors work, Wishart, Como, and Kidambi) in Arizona can be useful for high-temperature testing, especially of sensor performance and durability, but would be incredibly expensive to attempt to provide snow conditions that are possible at Mcity or ACM, and vice versa. A list of facilities with specific CAV testing capabilities is included in Wishart et al. (2020b). The International Alliance for Mobility Testing and Standardization is attempting to create a database of closed course test facilities including all of their respective capabilities so that CAV developers can find the facility that suits their testing needs quickly. In parallel, the AV Test Site Community of Practice under the SAE ORAD Committee is developing guidance documentation to closed course test facilities to ensure safe CAV testing takes place.

Closed course testing is incredibly useful, but the biggest disadvantage is that access to a test facility can be expensive, especially for smaller companies. Building a facility is very capital intensive: Mcity reportedly cost $10 million dollars (Gardner, 2015) while the ACM's first phase was $35 million dollars and the total expected cost is $135 million dollars (Walsh, 2020). Operating costs must also be factored in, and large facilities require funding for repairs and maintenance, personnel, taxes, utilities, etc. Cost, time, and access also impact which scenarios are chosen for validation of simulation results, which means that testing scenarios must be selected judiciously, or the investment will not be optimized. Like simulation testing, fidelity with the real world is an issue, as can be seen by the obvious differences in the surrogate human of Figure 8.9, although the fidelity is easier to quantify. These issues will be discussed further in section "Safety Case."

Public Road Testing

Real-world testing of vehicles on public roads is also known as Naturalistic-Field Operational Tests (N-FOTs). Public road testing of CAVs, depicted in Figure 8.10, is the most visible (to the public) of the CAV V&V process. Public road testing must involve careful planning to ensure that the ADS is only controlling the CAV within its defined ODD but the complete fidelity means that there can be high confidence in the results and obviously no need for subsequent validation. Just as simulation testing and closed course testing are connected for iterative validation, public road testing and simulation testing are iteratively conducted to inform the CAV developer's ODD (i.e., learning and improvement allows for expansion), but also to identify interesting scenarios that can then be examined more closely in simulation, parameterized to test for response robustness, and added to the scenario library. To accomplish this connection between test methods, an HD map must be created of the public road location and imported to the simulation tool. It should be noted that many CAV developers will only test on public roads where they have pre-existing HD maps.

FIGURE 8.10 Waymo CAV in public road testing.

Photo by Daniel Lu.

As mentioned previously, there are two different types of public road testing in shadow driving and safety driving. Shadow driving is safer in that the human driver is in control of the DDT and the ADS is operating in the background; however, since the trajectory determined by the path planning sub-system is not used, it is difficult to appreciate the actual impact that the trajectory would have had on the surrounding traffic participants. Safety driving, done with professional drivers, allows the ADS to be in complete control of the vehicle, with the safety driver performing the OEDR and ensuring that the ADS is operating safely. However, there are questions about whether humans can monitor automation very well (Hancock, Nourbakhsh, & Stewart, 2019), and accidents like the Tempe, AZ incident of 2018 could occur. As an aside, in October 2019, Waymo removed the safety driver from some of the CAVs that were being tested in Chandler, AZ to become the first robo-taxi service in the world.

The advantages of public road testing are obvious, such as testing in real-world conditions with the full CAV, potentially in an ODD that the CAV developer intends to deploy in commercialization. The U.S. DOT has also established testbeds for connected vehicles (but the equipment can be useful for CAV testing) that provide some of the advantages of a closed course facility in terms of off-board test equipment, with locations in Anthem, AZ (Maricopa County Department of Transportation, 2017); Palo Alto, CA; Oak Ridge, TN; Novi, MI; Mclean, VA; Manhattan, NY; and Orlando, Fl. It is difficult to imagine a V&V process that does not include a significant amount of public road testing. Indeed, several CAV developers have accumulated substantial mileage on public roads, with nearly 2 million miles accumulated in California in 2020 alone, and this was a decrease (due to the COVID-19 pandemic) of 29% from 2019 (Hawkins, 2021). However, as mentioned previously, the industry sees simulation testing as the method that should see the bulk of the effort (Kalra & Paddock, 2016).

There are some drawbacks to public road testing of CAVs. There is some controversy over testing in public due to the prototypical and nascent technological nature of the CAVs being tested and whether this is a danger to public safety (Hancock, Nourbakhsh, & Stewart, 2019). This controversy is likely heightened (at least temporarily) when incidents like the Tempe, AZ Uber accident or the multiple Tesla accidents occur. Public road testing thus presents a dilemma: it is essential to CAV development, but if it is not done safely (even by a minority of CAV developers), public trust erosion could mean government regulation that limits or even prohibits it.

Public road testing is also very expensive and time consuming. The expenses include the costs of the CAVs themselves, the operations costs (including personnel and fuel/electricity), and the repairs/maintenance costs. Public road testing is time consuming because a long period of time is needed in order to experience a wide variety of scenarios, including safety-critical scenarios that are key to ADS development. To provide some context, in the United States, a police-reported crash occurs every 530 thousand miles; a fatal crash occurs every 100 million miles (this distance is longer than the distance between the Earth and the sun). This means that a large proportion of the N-FOTs do not involve significant or "interesting" scenarios. Completing 100 million miles in a public road-testing program would take 100 vehicles traveling at 25 mph average speed of some 40 thousand hours to complete. The mileage accumulation will also be impacted by the complexity of the testing environment. CAV developer Cruise compared the instances of maneuvers and scenarios between San Francisco and the Phoenix suburbs and found the ratios of some instances (e.g., encountering emergency vehicles) to be as high as 46.6:1, but even left turns were more common (by a ratio of 1.6:1) in San Francisco. Cruise's position is that a minute of testing in San Francisco is equivalent in learning value to an hour of testing in the Phoenix suburbs. The following ratios from Cruise are for instances per 1,000 miles of safety driving in San Francisco and Phoenix (Vogt, 2017):

- Left turn 16:1
- Lane change 5.4:1
- Construction blocking lane 19.1:1
- Pass using opposing lane 24.3:1
- Construction navigation 39.4:1
- Emergency vehicle 46.6:1

Another public road-testing location consideration is that laws on public road testing vary by state, even though it can seem as though states compete to be the most CAV friendly. The IIHS keeps track of the laws and provisions for each state as well as the District of Columbia in a database that includes [Insurance Institute for Highway Safety (IIHS), 2021]

- Type of driving automation that is allowed:
  - In 12 states, CAV public road testing is legal.
  - In 15 states, CAV deployment is legal.
  - In three states, CAV deployment is legal but only for commercial vehicles.
  - One state (MI) authorizes testing of an "automated motor vehicle" and deployment of "on-demand automated motor vehicle networks", the latter meaning ride-hailing companies like Lyft or Uber.
  - One state's DOT (PA) has voluntary guidance that authorizes "highly automated vehicles", i.e., Level 4, while the law allows deployment of "highly automated work zone vehicles".
- Whether an operator is required to be licensed:
  - In six states, the operator does not need to be licensed.
  - In 11 states, the operator must be licensed.
  - In six states, the license requirement depends on the level of automation, mostly for Level 5 CAVs (FL, GA, NV, NC, and ND) and NE for Level 4 CAVs.
  - One state's DOT (PA) has voluntary guidance requiring a licensed operator for testing Level 4 CAVs, but the law authorizing "highly automated work zone vehicles" does not require the operator to be licensed.
  - In six states, the issue of operator licensing is not addressed in the existing legislation/provision.

- Whether an operator must be in the vehicles:
  - In 10 states, there is no requirement for an operator to be in the vehicle.
  - In seven states, there is a requirement for an operator to be in the vehicle.
  - In eight states, the requirement depends on the level of automation, mostly for Level 5 CAVs (AZ, FL, GA, NV, and ND) and IA and NE for Level 4 CAVs.
  - One state's DOT (PA) has voluntary guidance requiring an operator in Level 4 CAVs, but the law authorizing "highly automated work zone vehicles" does not require the operator inside the CAV.
  - One state (NH) only requires an operator if public road testing (rather than deployment) is occurring.
  - In five states, the issue of an operator inside the CAV is not addressed in the existing legislation/provision.
- Whether liability insurance is required:
  - In four states, no liability insurance is required.
  - In 24 states, liability insurance is required. In nine of those 24, minimum liability amounts are specified at either $2,000,000 or $5,000,000.
  - In three states, the issue of liability insurance is not addressed in the existing legislation/provision.

In addition to the patchwork of state laws governing CAV public road-testing location choice, CAV developers looking to establish a base of operations from which to conduct public road testing must carefully weigh their options, including a number of factors such as:

- Number of test vehicles to utilize
- Whether one or two safety drivers are present
- Level of complexity of the ODD
- Whether a location with additional equipment (such as the connected vehicle testbeds mentioned above) is desired
- Mileage accumulation rates
- Support staff and equipment (i.e., to service-disabled test vehicles), including training
- Test data storage and upload

The role of public road testing will be discussed further in section "Safety Case."

Evaluation Methods

The OSA metrics and test methods can produce results, but in order to be able to assess the operational safety of a CAV as it navigates a scenario, an evaluation methodology that allows for an interpretation of the metrics measurements is required, including providing context of the test methodology and scenario being used. While there has been research conducted on OSA metrics and test methods, as well as how to build the safety case from testing, to the authors' knowledge, there is no existing evaluation methodology in the literature. This presents a gap in the V&V process that must be addressed.

The same IAM group that proposed the OSA metrics of Table 1 is also developing what they have dubbed the OSA methodology. The OSA methodology is a way of scoring the navigation of a single, given scenario *S*. In the current form, the formulation includes quantified scenario relevance and scenario complexity in a test with quantified fidelity, with measurements of *n* relevant OSA metrics. The fidelity is quantified by the type of test method employed, most likely simulation, closed course, and public road testing (with the latter fidelity at 1). Scenario relevance refers to how reasonably foreseeable the scenario is to be expected by a CAV in the specified ODD. Scenario complexity is a quantification of how complex the scenario is, including speeds, number of traffic participants, road geometry, environmental conditions, etc. (e.g., as shown in section "Public Road Testing," the complexity of testing in San Francisco is higher, in general, than that of Phoenix suburbs). The current, proposed formulation of the OSA methodology is:

$$OSA_S = Fidelity_S \times Relevance_S \times Complexity_S \times \left(1 - \sum_{i=1}^{n} Severity_i \times OSA\ Metric\ Value_i\right) \quad (1)$$

In Equation (1), fidelity, relevance, and complexity all range from 0 to 1. The OSA metric values also range between 0 and 1 (inclusive; violations are binary, and errors are percentages). The severity parameter for each OSA metric violation or error ranges from 0.1 to 1, where larger values signify a more severe violation or error penalty. The maximum score that a CAV can achieve when navigating a given scenario is 1 while the lowest is 0.

The OSA methodology requires significant research to determine how to calculate each term, and the overall methodology must be validated using simulation and real-world data. Work has been done on some of the terms, namely, fidelity (Koopman & Wagner, 2018), and relevance and complexity (Feng, Feng, Yu, Zhang, & Liu, 2019), and a full literature review is being conducted. To determine the relevant OSA metrics, the following preliminary decision structure for Black Box testing has been proposed:

1. Select the test method and determine the test fidelity. If public road testing, proceed to Step 4.
2. Determine the behavioral competency to be tested.
3. Construct the scenario and determine the scenario relevance and complexity.
4. Conduct the test and obtain measurements of the OSA metrics.
5. Was the result of the test a collision incident? If "yes," then the OSA score is 0. The collision severity can also be determined if a more severe collision is to be penalized (i.e., a negative OSA score). If no, proceed to Step 7.
6. Were any rules of the road violated? If "yes," then calculate the severity of the violation.
7. Were there safety envelope metric(s) violations? If "yes," then calculate the severity of the violation(s).
8. Calculate the OSA score for the scenario navigation.

If Gray Box or White Box testing is being conducted, then more OSA metrics from Table 1 would be included in the decision structure. The objective will be to develop the methodology to a level of maturity and validation that it can be adopted and used by other organizations. For example, the V&V Task Force under the SAE ORAD Committee plans to develop an evaluation methodology document after the OSA metrics document is complete. It is possible that other research groups will be developing evaluation methodologies, and continuous surveying of the literature will be required.

Evaluation Criteria

Once the testing is complete and the OSA metrics measurements have been interpreted by the evaluation methodology, the final step is determining whether the level of operational safety is sufficient. This determination is often given the shorthand "How safe is safe enough?", and several research groups have conducted surveys of the general public to attempt to answer the question, for example, (Shariff, Bonnefon, & Rahwan, 2021) and (Liu, Yang, & Xu, 2019). The results indicate that the operational safety performance of CAVs may have to be significantly higher than for human-driven vehicles. This is logical enough, as humans are more likely to tolerate other humans failing than of CAVs that have technology, including inscrutable AI, that the general public is unlikely to fully understand. Just how much higher the operational safety performance must be for CAVs is unclear at present.

If the Test and Validation methods are used exclusively in which the system (i.e., the CAV) is tested in scenarios, one possible method for determining the level of operational safety, i.e., establishing the evaluation criteria, is to set a minimum OSA methodology for an individual scenario. The minimum score would likely depend on the complexity and relevance of the scenario, perhaps among other factors. The safety case would, as discussed in section "Safety Case," then be constructed using an aggregate of OSA methodology scores. If additional V&V methods are used, either component Test and Validation methods or Design and Development methods, the establishment of evaluation criteria will be required for each of these diverse elements of the safety case. It should be noted that this approach has been proposed by the authors (Wishart and Como) and is not yet an agreed-upon method for establishing evaluation criteria.

Before the evaluation criteria can be determined, benchmarking of the operational safety of human drivers is required. The RAND Corporation showed that in order for CAVs to match the operational safety of human drivers (e.g., 1.09 fatalities per 100 million miles) at a 95% confidence level, a CAV test fleet would need to accumulate 275 million miles without a collision incident involving a fatality. Figure 8.11 shows the number of failure-free CAV miles to match human driver operational safety performance in terms of collisions involving fatalities, collisions involving reported injuries (77 per 100 million miles), collisions involving an estimated total of injuries (103 per 100 million miles), reported collision incidents (190 per 100 million miles), and estimated total collision incidents (382 per 100 million miles) (Kalra & Paddock, 2016). The data from Figure 8.11 can serve as a starting point for discussion on the evaluation criteria.

The determination of the evaluation criteria has thus only just begun. This is not an engineering problem; although CAV industry representatives should absolutely be involved in the determination process, a wide variety of stakeholders must participate including representatives from government, academia, public safety organizations, and the general public. The diversity in stakeholders could make the determination more robust; however, the actual operational safety performance of CAVs, especially incidents that gain negative publicity such as high-profile accidents, could result in changing thresholds for CAV performance over time. It is also highly likely that the determination will be region and culturally dependent, as the results for the variations on the classic "trolley problem" found by the MIT Moral Machine researchers have shown [Massachusetts Institute of Technology (MIT), n.d.]. Finally, it is possible that different criteria will be established for CAV developers in

FIGURE 8.11 Failure-free CAV miles needed to demonstrate comparable operational safety performance to human drivers.

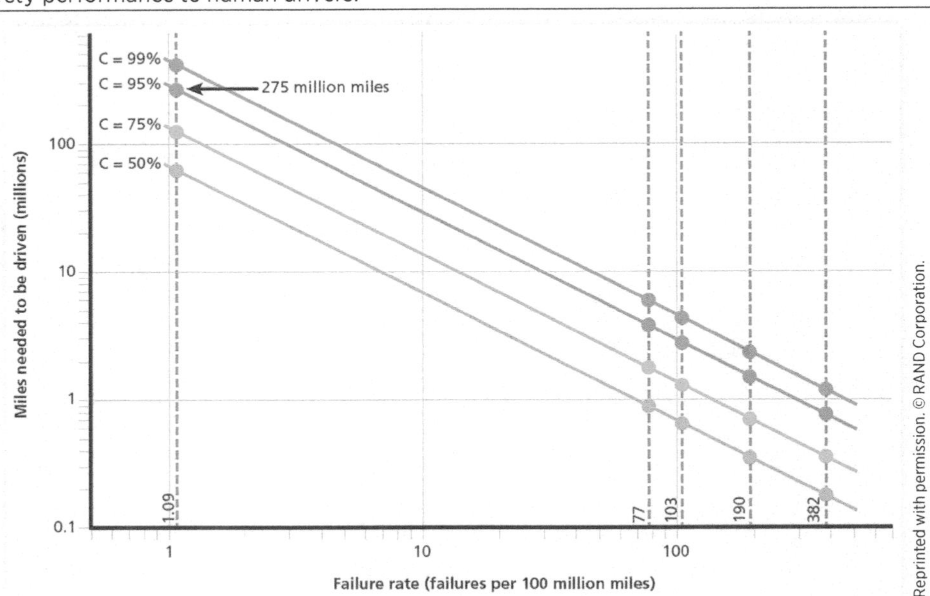

order to first be allowed to conduct public road testing with CAV prototypes and then be allowed to deploy commercial versions of their CAV.

Safety Case

One proposed deliverable from the V&V process is the safety case. The safety case is defined as "A structured argument, supported by a body of evidence, that provides a compelling, comprehensible, and valid case that a product is safe for a given application in a given environment" (United Kingdom Ministry of Defense, 2007). As was described earlier in this chapter, there is no regulation for a V&V process, and the same applies to how to construct a safety case. The closest guidance that CAV developers have at present is the VSSA guidance from the NHTSA and the Underwriters Laboratories (UL) 4600 standard.

The VSSA guidance from the NHTSA AV Policy 2.0 document (U.S. Department of Transportation—National Highway Traffic Safety Administration, 2017) was released in September 2017 and contains 12 design principles that a CAV developer could include in their safety case. The design principles are not highly detailed, and no specific data are recommended to be included in the safety case.

UL 4600 was released in April 2020 and includes safety principles and a framework for evaluating a product like a CAV. The framework also includes risk assessment principles, human-machine interaction, AI, design processes, and testing and validation and helps a CAV developer with the elements of a safety case [Underwriters Laboratories (UL), 2020]. The UL 4600 standard does not provide prescriptive steps for safety case construction, as it provides a framework only; however, safety of a product is never assumed and must be proven. The UL framework considers work products from accepted standards like the Automotive Safety Integrity Level (ASIL) concept of ISO 26262 [International Organization for Standardization (ISO), 2018] and the scenario classification taxonomy (1) known safe,

(2) known unsafe, (3) unknown safe, and (4) unknown unsafe from ISO/PAS 21448 [International Organization for Standardization (ISO), 2019]. The current consensus in the CAV community is that ISO 26262 and ISO/PAS 21448 are important components of a CAV safety case. UL 4600 does not prescribe the evaluation methods or criteria of the previous sections of this chapter. Finally, UL 4600 requires either an internal auditor or a third-party auditor. The UL 4600 standard is just one way to construct a safety case, but despite these exclusions, the standard is widely seen as a positive development toward V&V standardization.

The safety case can be constructed, in part, using any or all of the V&V methods of section "Challenges"; as noted above, any safety case is likely to include concepts from ISO 26262 and ISO/PAS 21448. Other safety case elements could include following best practices in safety culture, change management, and training. It should be noted that the safety case must also include ensuring that the vehicle meets the FMVSS, which is regulated by the NHTSA. The safety case is also likely to include V&V methods for the ADS hardware and ADS software as well as the overall vehicle (i.e., the system).

The Test and Validation elements of the safety case construction, and the three main test methods, in particular, will be the focus of the rest of this section. It should not be forgotten, however, that the Design and Development methods are likely to be employed by a CAV developer as well in their V&V process and by regulations. A safety case constructed from simulation testing, closed course testing, and public road testing, and incorporating the OSA methodology concept, is presented.

The process flowchart of the safety case construction from testing is shown in Figure 8.12. The steps of this process are:

1. Determine the ODD of the CAV (by the CAV developer).
2. Based on this ODD, a scenario library can be generated or obtained from a third party. The generation can occur by public road testing (either shadow or safety driving) the subject CAV. A by-product of the public road testing, in addition to generating scenarios, is the OSA results ("OSA Results 3"), which are inputs that modify the safety case status.
3. Using the scenario library, a scenario for simulation testing is selected and the simulation testing is conducted. The OSA results ("OSA Results 1") are inputs that modify the safety case status and also the ODD. More simulations can be conducted iteratively. Alternatively, the scenario(s) and OSA results are input to the closed course testing.

FIGURE 8.12 Safety case constructed from simulation, closed course, and public road testing.

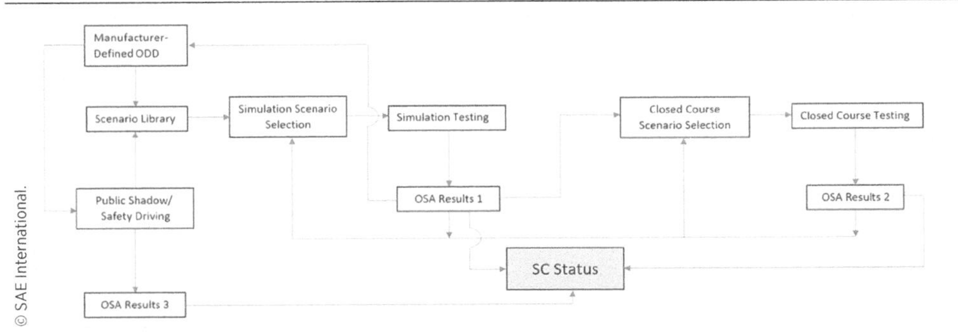

4. A scenario for closed course testing is selected and the closed course testing is conducted. The OSA Results ("OSA Results 2") are inputs that modify the safety case status and also inform the simulation scenario selection by returning to Step 3 for iteration of simulation testing.
5. Additional public road testing can take place (in parallel to the simulation and closed course testing in some cases) to continue to add to the scenario library, update the safety case status, or both.

The safety case status is constantly updated for all test scenarios for which there are OSA results:

$$SC\ Status = \sum_{0}^{All\ Test\ Scenarios} OSA\ Results\,1 + OSA\ Results\,2 + OSA\ Results\,3 \quad (2)$$

The safety case status threshold above which the CAV is deemed ready for a given deployment will be determined by the evaluation criteria discussed in section "Evaluation Criteria." It should be noted that the safety case is dynamic: any testing will alter the safety case status, increasing the aggregate score by an OSA result that is above the minimum threshold for that particular scenario and decreasing the aggregate score by an OSA result that is below the minimum threshold for that particular scenario.

The next consideration in test-and-validation-based V&V and safety case construction is ensuring that sufficient testing has been completed that is both representative of the ODD but has also sufficiently sampled the unusual scenarios, identified here as corner cases, edge cases, and long-tail cases [adapted from (SAE International and Deutsches Institut für Normung, 2019] and (SAE International, In preparation); note that "capabilities" in these definitions refers specifically to "DDT performance capabilities" and "parameter values" in this context refer to vehicle motion control parameters values):

- Corner case: The combination of two or more parameter values, each within the capabilities of the ADS, but together constitute a condition that challenges its capabilities. A corner case may also be a condition for which the ADS response is not robust, i.e., a perturbation can cause a significant change in response.
- Edge case: Extreme parameter values or even the very presence of one or more parameters that result(s) in a condition that challenges the capabilities of the ADS. An edge case may also be a condition for which the ADS response is not robust, i.e., a perturbation can cause a significant change in response.
- Long-tail case: A rare condition that challenges the capabilities of the ADS. A condition could be exogenous such as a test scene or scenario, or it could be endogenous such as a hardware failure.

It should be noted that these definitions are not universally used, and the three are often conflated and labeled using (usually) "edge." These unusual cases may or may not be the same among CAV developers, although some common difficult scenarios include so-called zipper merges and construction zones, and Waymo identified the following scenarios as edge cases in their VSSA (Waymo, 2017):

- Flashing yellow arrow signals
- Wrong-way drivers

- Cyclists making left turns
- Pedestrians, especially those that are "zig-zagging" while crossing the street
- Motorcycles "splitting the lane"

One way of determining whether a particular scenario is an edge case or corner case is to first analyze the ADS behavior and determine whether the behavior is rooted in a correct, logical understanding of the scenario. If "no," then the scenario might be an edge/corner case. If the behavior does originate from a correct, logical understanding of the scenario, but the behavior is not robust to a perturbation, then the scenario might still be an edge/corner case. As an example, a "base scenario" can be perturbed and reveal an edge/corner case, as shown in Figure 8.13. The first perturbation is the removal of the center lane; the second is an unknown class of object (or if the AI has been poorly trained); the third perturbation has a known class of object, but also an unfavorable environmental condition (the depicted lightning).

FIGURE 8.13 Base scenario and perturbations resulting in edge/corner cases.

In general, the traffic participants that can contribute to a scenario being an edge/corner/long-tail case can be categorized as vehicle type, infrastructure type, and VRUs, with some examples of these shown in Table 8.2.

TABLE 8.2 Traffic participants that can contribute to a scenario being a corner/edge/long-tail case.

Vehicle type	Infrastructure type	Vulnerable road users
Emergency vehicles	Special zones	Crowds
Law enforcement vehicles	Maintenance	Cyclists
Transit vehicles	Unexpected objects	Animals
Commercial vehicles	Parking lot	Other
Construction/Utility vehicles	Sudden inclement weather	
Farm/Large payload vehicles	Non-mapped areas	

The possible scenarios within any ODD are effectively infinite in number, and so the concept of coverage-driven test and validation is adopted from the software testing industry (Tahir & Alexander, 2020). Coverage-driven test and validation requires a machine-readable scenario description language and a tool to develop scenario variants, track which scenarios have been tested, and determine which scenarios should be tested next and with what test method (Foretellix, n.d.). The techniques used for these tools include random number generation, search-based software testing, machine learning, and high-throughput testing (Tahir & Alexander, 2020).Another technique taken from ISO 21448 – Safety of the Intended

FIGURE 8.14 Scenario areas characterized by known and unknown CAV navigation.

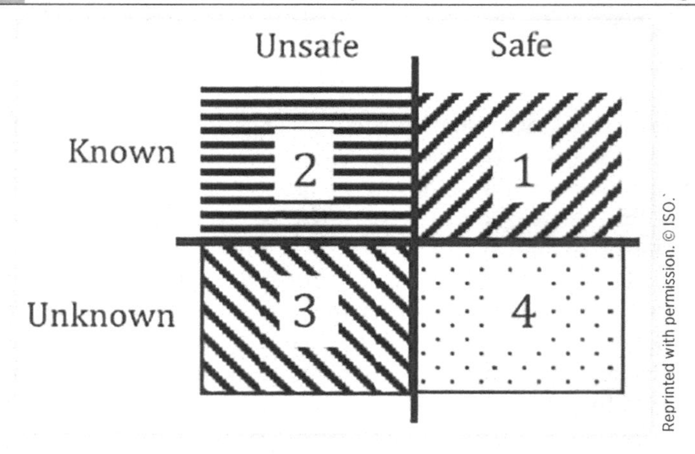

Functionality (SOTIF) [International Organization for Standardization (ISO), 2019] (and mentioned earlier in the chapter) is to classify scenarios into one of four areas shown in Figure 8.14 based on whether the scenario is difficult for the CAV to navigate:

1. Known safe
2. Known unsafe
3. Unknown unsafe
4. Unknown safe

At the beginning of the CAV development, the scenario space might look like the left-hand side of Figure 8.15, with large areas of known unsafe scenarios and unknown unsafe

FIGURE 8.15 Scenario classification technique example.

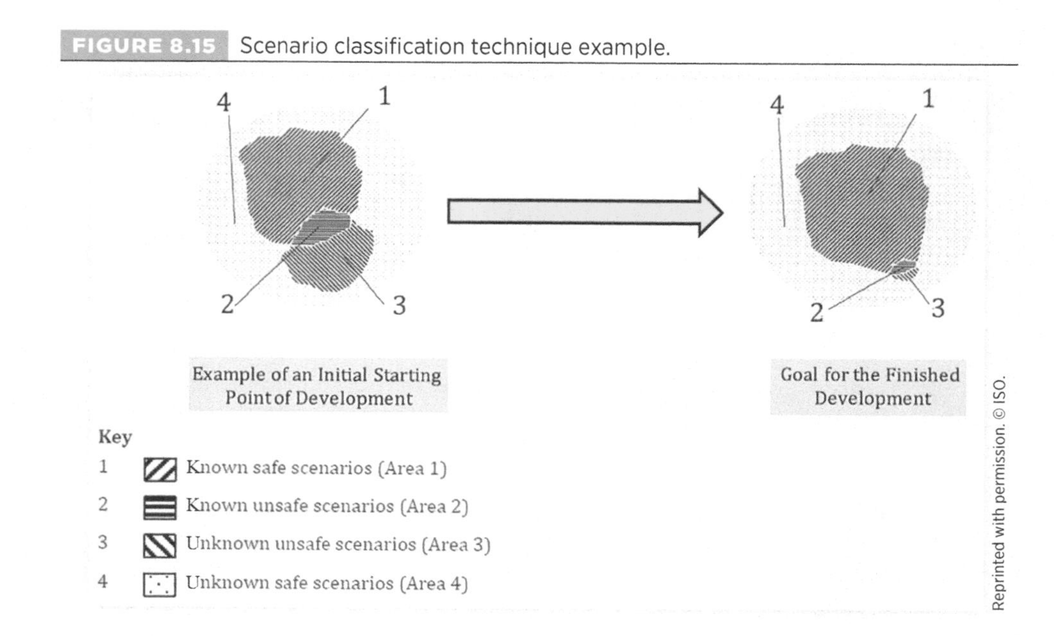

scenarios. The objective is to evaluate the ADS performance in Area 2 and Area 3 and prove that these areas are sufficiently small, i.e., that the risk is sufficiently small, as illustrated in the right-hand side of Figure 8.15. Demonstration that the risk is acceptable is a way of constructing the safety case, the culmination of the V&V process.

References

Ahamed, M., Tewolde, G., & Kwon, J. (2018). Software-in-the-Loop Modeling and Simulation Framework for Autonomous Vehicles. *IEEE International Conference on Electro/Information Technology (EIT)*. Rochester, MI. doi:10.1109/EIT.2018.8500101

ASAM. (2021, March 4). ASAM OpenDRIVE, Version 1.6.1.

Automated Vehicle Safety Consortium (AVSC). (2020). Best Practices for Safety Operator Selection, Training, and Oversight Procedures for Automated Vehicles under Test.

Automated Vehicle Safety Consortium. (2021, March). Best Practice for Metrics and Methods for Assessing Safety Performance of Automated Driving Systems (ADS). SAE Industry Technologies Consortium.

Do, W., Rouhani, O., & Miranda-Moreno, L. (2019, January 26). Simulation-Based Connected and Automated Vehicle Models on Highway Sections: A Literature Review. *Journal of Advanced Transportation (Special Issue: Partially Connected and Automated Traffic Operations in Road Transportation)*, 2019, 1-14. doi:10.1155/2019/9343705

Easterbrook, S. (2010, November 29). The Difference between Verification and Validation. Retrieved from easterbrook.ca: www.easterbrook.ca/steve/2010/11/the-difference-between-verification-and-validation/

Elli, M., Wishart, J., Como, S., Dhakshinamoorthy, S., & Weast, J. (2021). Evaluation of Operational Safety Assessment (OSA) Metrics for Automated Vehicles in Simulation. SAE Technical Paper 2021-01-0868. https://doi.org/10.4271/2021-01-0868

Fadaie, J. (2019). The State of Modeling, Simulation, and Data Utilization within Industry—An Autonomous Vehicles Perspective.

Feng, S., Feng, Y., Yu, C., Zhang, Y., & Liu, H. (2019). Testing Scenario Library Generation for Connected and Automated Vehicles, Part I: Methodology, *IEEE Transactions on Intelligent Transportation Systems*, 22(3), 1573-1582. doi:10.1109/TITS.2020.2972211

Foretellix. (n.d.). Foretify. Retrieved from https://www.foretellix.com/technology/

Gardner, G. (2015, July 20). U-M Opens $10M Test City for Driverless Vehicle Research. Retrieved from Detroit Free Press: https://www.freep.com/story/money/cars/2015/07/20/university-michigan-mdot-general-motors-ford-toyota-honda-nissan/30415753/

Hancock, P., Nourbakhsh, I., & Stewart, J. (2019). On the Future of Transportation in an Era of Automated and Autonomous Vehicles. *Proceedings of the National Academy of Sciences of the United States of America (PNAS)*, 116(16), 7684-7691.

Hawkins, A. (2021, February 11). Waymo and Cruise Dominated Autonomous Testing in California in the First Year of the Pandemic. Retrieved from theverge.com: https://www.theverge.com/2021/2/11/22276851/california-self-driving-autonomous-cars-miles-waymo-cruise-2020

Hyperdrive. (2020, May 15). The State of the Self-Driving Car Race 2020. Retrieved from Bloomberg.com: https://www.bloomberg.com/features/2020-self-driving-car-race/

IEEE. (2021). Standard for Assumptions for Models in Safety-Related Automated Vehicle Behavior.

Insurance Institute for Highway Safety (IIHS). (2021, May). Autonomous Vehicle Laws. Retrieved from iihs.org: https://www.iihs.org/topics/advanced-driver-assistance/autonomous-vehicle-laws

International Organization for Standardization (ISO). (2018). ISO 26262-1:2011(en), Road Vehicles—Functional Safety—Part 1: Vocabulary. Retrieved October 17, 2018, from https://www.iso.org/obp/ui/#iso:std:iso:26262:-1:ed-1:v1:en

International Organization for Standardization (ISO). (2019). ISO/PAS 21448:2019—Road Vehicles—Safety of the Intended Functionality.

International Standards Organization (ISO). (2020). ISO/TR 4804:2020—Road Vehicles—Safety and Cybersecurity for Automated Driving Systems—Design, Verification and Validation.

Kalra, N., & Paddock, S. (2016). Driving to Safety: How Many Miles of Driving Would It Take to Demonstrate Autonomous Vehicle Reliability? Rand Report.

Koopman, P., & Wagner, M. (2016). Challenges in Autonomous Vehicle Testing and Validation. SAE Technical Paper 2016-01-0128. https://doi.org/10.4271/2016-01-0128

Koopman, W., & Wagner, M. (2018). Toward a Framework for Highly Automated Vehicle Safety Validation. SAE Technical Paper 2018-01-1071. https://doi.org/10.4271/2018-01-1071

Liu, P., Yang, R., & Xu, Z. (2019). How Safe Is Safe Enough for Self-Driving Vehicles? *Risk Analysis*, 39(2), 315-325.

Ma, J., Zhou, F., Huang, Z., & James, R. (2018). Hardware-in-the-Loop Testing of Connected and Automated Vehicle Applications: A Use Case For Cooperative Adaptive Cruise Control. *21st International Conference on Intelligent Transportation Systems (ITSC)*. Maui, HI, 2878-2883. doi:10.1109/ITSC.2018.8569753

Mahmud, S., Ferreira, L., Hoque, M., & Tavassoli, A. (2017). Application of Proximal Surrogate Indicators for Safety Evaluation: A Review of Recent Developments and Research Needs. *IATSS Research*, 41(4), 153-163.

Maricopa County Department of Transportation. (2017). Connected Vehicle Test Bed: Summary.

Massachusetts Institute of Technology (MIT). (n.d.). Moral Machine. Retrieved from https://www.moralmachine.net/

National Traffic Safety Board (NTSB). (2019). Collision Between Vehicle Controlled by Developmental Automated Driving System and Pedestrian, Tempe, Arizona, March 18, 2018. Accident Report—NTSB/HAR-19/03—PB2019-101402.

Omidvar, A., Pourmehrab, M., & Emami, P. (2018). Deployment and Testing of Optimized Autonomous and Connected Vehicle Trajectories at a Closed-Course Signalized Intersection. *Transportation Research Record: Journal of the Transportation Research Board*, 2672(19), 45-54. doi:10.1177/0361198118782798

Park, C., Chung, S., & Lee, H. (2020, April). Vehicle-in-the-Loop in Global Coordinates for Advanced Driver Assistance System. *Applied Sciences*, 10(8), 2645. Retrieved from https://doi.org/10.3390/app10082645

Paukert, C. (2016, January 19). Mcity: America's True Nexus of Self-Driving Research. Retrieved from cnet.com: https://www.cnet.com/roadshow/news/mcity-americas-true-nexus-of-self-driving-research/

Plummer, A. (2006). Model-in-the-Loop Testing. *Proceedings of the Institution of Mechanical Engineers, Part 1 - Journal of Systems and Control Engineering*, 229(3), 183-199.

SAE International Surface Vehicle Recommended Practice, "Safety-Relevant Guidance for On-Road Testing of Prototype Automated Driving System (ADS)-Operated Vehicles," SAE Standard J3018, Revised December 2020.

SAE International Surface Vehicle Recommended Practice, "Definitions for Terms Related to Automated Driving Systems Reference Architecture," SAE Standard J3131, In preparation.

SAE International and Deutsches Institut für Normung. (2019, June). Terms and Definitions Related to Testing of Automated Vehicle Technologies. DIN SAE SPEC 91381.

SAE International Surface Vehicle Information Report, "Taxonomy and Definitions of Terms Related to Verification and Validation of ADS," SAE Standard J3208, In preparation.

Schmidt, B. (2021, April 27). No Need for Radar: Tesla's Big Jump forward in Self-Driving Technology. Retrieved from The Driven.com: https://thedriven.io/2021/04/27/no-need-for-radar-teslas-big-jump-forward-in-self-driving-technology/

Shariff, A., Bonnefon, J.-F., & Rahwan, I. (2021). How Safe Is Safe Enough? Psychological Mechanisms Underlying Extreme Safety Demands for Self-Driving Cars. *Transportation Research Part C: Emerging Technologies*, 126, 103069. doi:doi.org/10.1016/j.trc.2021.103069

Siddiqui, F. (2021, May 14). Tesla Is Like an 'iPhone on Wheels.' And Consumers Are Locked into Its Ecosystem. Retrieved from washingtonpost.com: https://www.washingtonpost.com/technology/2021/05/14/tesla-apple-tech/

Tahir, Z., & Alexander, R. (2020). Coverage Based Testing for V&V and Safety Assurance of Self-Driving Autonomous Vehicle: A Systematic Literature Review. *The Second IEEE International Conference on Artificial Intelligence Testing*. Oxford, UK. doi:10.1109/AITEST49225.2020.00011

Tettamanti, T., Szalai, M., Vass, S., & Tihanyi, V. (2018). Vehicle-in-the-Loop Test Environment for Autonomous Driving with Microscopic Traffic Simulation. *IEEE International Conference on Vehicular Electronics and Safety (ICVES)*. Madrid, Spain. doi:10.1109/ICVES.2018.8519486

U.S. Department of Transportation—National Highway Traffic Safety Administration. (2017). Automated Driving Systems: A Vision for Safety 2.0.

Underwriters Laboratories (UL). (2020). UL 4600—Standard for Safety for the Evaluation of Autonomous Products.

United Kingdom Ministry of Defense. (2007). Defence Standard 00-56 Issue 4: Safety Management Requirements for Defence.

United Nations Economic Commission for Europe. (2020). ECE/TRAN/WP.29—Framework Document on Automated Vehicles.

University of California PATH Program. (2016, February). Peer Review of Behavioral Competencies for AVs. Retrieved from https://www.nspe.org/sites/default/files/resources/pdfs/Peer-Review-Report-IntgratedV2.pdf

USDOT—NHTSA. (2020, November). Framework for Automated Driving System Safety—49 CFR Part 571—Docket No. NHTSA-2020-0106.

Vogt, K. (2017, October 3). Why Testing Self-Driving Cars in SF Is Challenging but Necessary. Retrieved from medium.com: https://medium.com/cruise/why-testing-self-driving-cars-in-sf-is-challenging-but-necessary-77dbe8345927

Walsh, D. (2020, January 21). American Center for Mobility Opens Technology Park for Development. Retrieved from crainsdetroit.com: https://www.crainsdetroit.com/mobility/american-center-mobility-opens-technology-park-development

Waymo. (2017). On the Road to Fully Self-Driving. Waymo Safety Report to NHTSA.

Waymo. (2020, April 28). Off Road, But Not Offline: How Simulation Helps Advance Our Waymo Driver. Retrieved from Waymo.com: https://blog.waymo.com/2020/04/off-road-but-not-offline--simulation27.html#:~:text=To%20date%2C%20we%20have%20driven,the%20velocity%20of%20our%20learning.

Wikimedia Commons. (n.d.). Systems Engineering Process II. Retrieved from https://commons.wikimedia.org/wiki/File:Systems_Engineering_Process_II.svg

Wishart, J., Como, S., Elli, M., Russo, B., Weast, J., Altekar, N., … Chen, Y. (2020a). Driving Safety Performance Assessment Metrics for ADS-Equipped Vehicles. SAE Technical Paper 2020-01-1206. doi:https://doi.org/10.4271/2020-01-1206

Wishart, J., Como, S., Forgione, U., Weast, J., Weston, L., Smart, A., … Ramesh, S. (2020b). Literature Review of Verification and Validation Activities of Automated Driving Systems. *SAE International Journal of Connected and Automated Vehicles*, 3(4), 267-323. doi:https://doi.org/10.4271/12-03-04-0020

Xu, Z., Wang, M., Zhang, F., Jin, S., Zhang, J., & Zhao, X. (2017). PaTAVTT: A Hardware-in-the-Loop Scaled Platform for Testing Autonomous Vehicle Trajectory Tracking. *Journal of Advanced Transportation, Advances in Modelling Connected and Automated Vehicles*, 2017, 1-11. doi:10.1155/2017/9203251

VOL
MAP
RADIO
MEDIA
NAV
FRONT
REAR
PASSENGER
AIR BAG
A/C
CLIMATE
MODE
DUAL
CleanAir
TEMP
AUTO
TEMP

9

Outlook

As described in Chapter 1, the concept of CAVs and their potential to transform the transportation landscape have been discussed for decades, from science fiction works in the 1930s to GM's experimentation with roadway-embedded guidance in the 1950s to Ernst Dickmanns' automated Mercedes van in the 1980s to the sustained development that followed the DARPA Grand and Urban challenge competitions of 2001 and 2004. Even with the recent breakthroughs in automation and connectivity technology, significant work is needed prior to the commercialization of fully automated and connected vehicles on public roadways.

Throughout this text, the various fundamental concepts comprising the CAV system have been discussed in detail. The technological aspects of the vehicle functionality have been the primary focus to this point, describing the hardware and software needed to achieve the tasks of localization, connectivity, perception, and path planning; the sensor suites of hardware have been reviewed, and the architecture and developments of sensor fusion and computer vision have been explored. The previous chapters have provided the context needed to understand the past and present states of the CAV industry with a deeper dive into functionality; however, this chapter will discuss the current status of CAVs with regard to the technology adopted by the industry, the public road deployments will be explored, and the standards, regulations, and supporting activities currently guiding the industry will be summarized. It is difficult to predict the near-term future of the CAV industry; however, one certainty is the need for regulation and societal trust to unite manufacturers on a path toward a safer transportation future.

State of the Industry—Technology

As the development of CAVs continues, vehicle designs proposed by CAV manufacturers are evolving to reflect changes to projected user preferences, hardware and software capabilities, advances in Machine Learning (ML) and AI, and expectations for CAV deployment timelines. In 2015, Waymo's Firefly, shown in Figure 9.1 (top), took the first automated trip on public roadways in a vehicle that had no steering wheel, no brake or accelerator pedals, or any other driver inputs for that matter. It looks unlike most vehicles on the road and its cute, diminutive appearance introduced the public to a friendly, non-threatening vision of future mobility. Five years and approximately 20 million automated-driving miles later, the Chrysler Pacifica and all-electric Jaguar I-Pace are the sleeker but more conventional new bodies housing Waymo's ADS technology in 2021. While the basic components and hardware are similar, the difference in vehicle design between the two eras depicted in Figure 9.1 offers some insight into the current industry outlook. Expectations on the pace of CAV deployment have been somewhat tempered, and many technical challenges remain. As CAVs learn to navigate new environments and situations, they may rely on the guidance and intervention of human drivers who still need to manipulate traditional vehicle operator controls such as a steering wheel and pedals. At the same time, CAV developers, investors, and transportation agencies have recognized that for technical and economic reasons, some sectors, such as long-haul trucking and first-mile/last-mile connections, may see CAV penetration before others.

FIGURE 9.1 Early Waymo autonomous driving technology Firefly (top); Current Waymo fleet (from left) Chrysler Pacifica PHEV and Jaguar I-Pace for Waymo One robo-taxi service and the Class 8 long-haul truck for Waymo Via (bottom).

Source: Waymo

For automation, many of the current CAV developers agree on the use of sensor suites comprised of some combination of cameras, LIDAR, and RADAR units varying in range, position, and number. These units all continue to improve; Waymo's fifth-generation design touts 360° cameras capable of identifying objects at greater than 500 meters away (Hawkins, 2020). Similarly, LIDAR units have improved significantly over the years. Today's solid-state LIDARs now remove the need for mechanical moving parts, allowing these units to last longer with greater reliability and have a smaller size. Solid-state LIDAR units also provide advantages in performance although the FOV and potential eye-safety concerns are limitations to the design (Aalerud, Dybedal, & Subedi, 2020). As these units continue to improve and decrease in cost, they facilitate advanced perception in CAVs. While many OEMs are taking advantage of improved LIDAR capabilities and have bought into the need for LIDAR to sufficiently complete the DDT, Tesla is a notable outlier. It has denounced the technology as a hindrance toward the progression of CAVs with CEO Elon Musk famously stating, "LIDAR is a fool's errand...and anyone relying on LIDAR is doomed" (Musk, 2019). Going even further, in May 2021, Tesla announced that the perception systems for new Model 3 and Model Y vehicles being built would no longer contain RADAR units (although curiously, the transition to "Tesla Vision" for Model X and Model S would occur later) (Tesla, n.d.). It is clear that the industry is far from united on the enabling hardware suite for CAVs, and the supporting technology continues to evolve rapidly.

The picture for connectivity technology is murky as well. As discussed in Chapter 3, the winner between the competing technologies of DSRC and C-V2X has not yet been decided, and so both CAV developers and infrastructure owners have been reluctant to spend capital on the deployment of any one type for fear of future stranded assets. Further complicating the situation, the role of connectivity is not yet clear, as there is reluctance on the part of CAV developers to necessarily trust the information provided to them by outside entities for safety-critical decisions. A future of widespread V2X and even CDA could be incredibly beneficial in several areas, but the timing and how it might transpire is still unclear.

AI and ML have made major strides in recent years; however, looking into the future of CAVs, one may wonder what aspects have been revealed as not "real" intelligence by the recent decade of tremendous progress in AI through the magic of deep learning. In recent scientific publications and widely quoted media reports by prominent AI researchers, such aspects have been pointed out and several sobering messages have been communicated:

1. *Real intelligence is not merely pattern memorizing and retrieving from databases.* Statistical machine learning helps to identify patterns within datasets and thus tries to make predictions based on existing data. However, real intelligence is much more than pattern recognition and requires complex and compositional techniques for decision-making. While pattern recognition is a critical component of intelligence, it is not the only component.
2. *Real intelligence is beyond feed-forward neural networks with supervision.* Recent successes with supervised learning in feed-forward deep networks have led to a proliferation of applications where large, annotated datasets are available. However, humans commonly make subconscious predictions about outcomes in the physical world and are surprised by the unexpected. Self-supervised learning, in which the goal of learning is to predict the future output from other data streams is a promising direction, but much more work is needed.
3. *Real intelligence is not a model which excels at "independent and identically distributed" (i.i.d.) tests but struggles to generalize toward out-of-distribution (OOD)*

samples. The availability of large-scale datasets has enabled the use of statistical machine learning and has led to significant advances. However, the commonly used evaluation criterion includes the evaluation of the performance of models on test samples drawn from the same distribution as the training dataset, and it encourages winning models to utilize spurious correlations and priors in datasets under the i.i.d. setting. Studies have shown that training under this i.i.d. setting can drive decision-making to be highly influenced by dataset biases and spurious correlations. As such, evaluation of OOD samples has emerged as a metric for generalization.

4. *Real intelligence should not just passively accept the training data provided but should also actively acquire data*. An intelligent agent is an active perceiver if it knows why it wishes to sense and then chooses what to perceive, in addition to determining how, when, and where to achieve that perception. This remark supports the conclusion that methods beyond supervised learning, such as self-supervision and weak supervision coupled with data synthesis strategies, as well as test-time adaptation could be the pathway toward a "post-dataset era." In other words, real intelligence is task centric rather than dataset centric.
5. *Real intelligence is not brittle about common-sense reasoning*. Real intelligence can incorporate common-sense reasoning of various kinds, including hypothetical reasoning about actions in the world. It may be noted that reasoning about actions plays a crucial role in common-sense reasoning.

While many high-profile efforts have been proposed to address one or more of the aforementioned observations, many of them, unfortunately, still operate at the theoretical stage and are far from providing practical solutions to real-world problems. Similar sentiments are also expressed by others in AI literature as well as in popular media. Researchers in both academia and industry are taking initial, yet bold, steps toward addressing these recognized shortcomings of AI by forming practical evaluation protocols and developing more useful computational tools.

State of the Industry—Deployments

Similar to the CAV technological growth, the development of potential CAV applications has also been rapidly evolving as the industry progresses. Five principal categories for which applications of CAV technologies are currently being considered include:

1. Delivery—Removing human drivers from delivery vehicles offers cost savings in the form of reduced human driver hours and obviates the need for occupant safety measures. As an example, Nuro received the first temporary exemption from the U.S. DOT to remove components traditionally designed to protect occupants and assist the driver such as side-view mirrors and a windshield. Since these low-speed delivery vehicles are not intended to carry passengers, many existing rules surrounding vehicle safety become less relevant. Some entities are exploring the use of smaller CAVs on sidewalks or bike lanes, which reduces conflicts with vehicles but introduces new bike and pedestrian safety considerations.
2. Last-mile connections—CAV shuttles and circulators can offer effective last-mile connections between transportation hubs, such as train stations, transit centers, and large parking structures, to destinations in the vicinity that are at distances

too far to cover comfortably on foot. As of 2021, there are already examples of this type of service being offered in neighborhoods, business parks, and university campuses throughout the world. Since they usually traverse a set of defined routes within a constrained, geographic area, and often at low speeds, the localization and perception tasks for CAVs operating as last-mile connectors are simplified. Companies developing vehicles specifically designed for these types of services include Local Motors, EasyMile, May Mobility, and Navya.

3. Robo-taxis—Like last-mile connections, robo-taxis will be owned by larger organizations as part of a fleet, rather than as privately owned vehicles. However, they offer point-to-point transportation over larger distances, like most privately owned vehicles or taxis on the road today. While CAVs remain prohibitively expensive for most private ownership, they may be economically feasible as taxis and ride-hailing vehicles as they eliminate labor costs associated with human drivers and are able to be driven 24/7, stopping only to be refueled (or recharged, depending on the powertrain). As of 2021, Waymo One is perhaps one of the most publicly known robo-taxi operators, offering limited automated taxi services without a safety driver in the vehicle within the Phoenix, AZ metropolitan area (not all Waymo One vehicles have no safety driver).
4. Long-Haul Trucking—Due to the relatively simple, well-defined freeway environments that characterize most long-haul trucking routes, many analysts and investors see this as among the most viable applications of CAV technologies in the near term. In long-haul trucking, long stretches of flat, straight roadways may be traveled for days on end, resulting in driver fatigue, inattentiveness, and unfamiliarity with roadways. Human recognition and decision errors influenced by these factors are the leading cause of roadway crashes (NHTSA, 2015). However, these same conditions are conducive to automated driving. In addition to several of the well-known, traditional trucking OEMs, some robotics companies currently operating in the automated trucking industry include Waymo Via, TuSimple, and Embark. Locomation is another automated long-haul trucking company that also incorporates the concept of platooning.
5. Personally Owned—Likely the last use case to become populated with deployed CAVs is the personally owned CAV. Much of the current focus throughout the industry operates around fleet-based and commercially owned CAV solutions. It will likely take significant developments in legislation, performance, cost, and public acceptance for personally owned CAVs to become a reality. Short of the additional challenges, Tesla has made some of the boldest implementations to date, preparing many of their vehicles being sold today for automated driving capability with OTA software updates once they become available. GM CEO Mary Barra claimed in May of 2021 that she sees GM selling personal CAVs by the end of the decade (Korosec, 2021). Many claims of this nature have been made surrounding the timeline of personal CAVs; however, for economic, technical, and regulatory reasons, the future of these vehicles seems much less clear than the other categories described.

These categories include the prominent approaches of the industry in 2021 with no definite answer as to the order in which market growth will occur. One detailed study (Litman, 2021) analyzed the industry and proposed predictions for CAV market penetration as depicted in Figure 9.2. However, the true future of CAV deployment is unknown

FIGURE 9.2 CAV market penetration predictions.

Stage	Decade	New Sales	Fleet	Travel
Development and testing	2020s	0%	0%	0%
Available with large price premium	2030s	2-5%	1-2%	1-4%
Available with moderate price premium	2040s	20-40%	10-20%	10-30%
Available with minimal price premium	2050s	40-60%	20-40%	30-50%
Standard feature included on most new vehicles	2060s	80-100%	40-60%	50-80%
Saturation (everybody who wants it has it)	2070s	?	?	?
Required for all new and operating vehicles	?	100%	100%	100%

Autonomous vehicle will probably take several decades to penetrate new vehicle sales, fleets and travel.

and other studies show more optimistic or pessimistic projections. It is apparent by all of the overly optimistic predictions of complete CAV commercialization and even the overly pessimistic predictions of CAV technology being a thing of science fiction that nobody truly knows what the future holds for CAVs, but what should be made clear from this book is that much effort has gone into the development of CAV technology with an end goal of contributing to an overall safer transportation network worldwide.

Since around 2015, it has seemed that OEMs and start-ups have consistently declared they will be selling full CAVs within the next couple of years. However, as of 2021, it is still not possible to visit a dealership and purchase a vehicle with Level 5, Level 4, or even Level 3 automation (Audi canceled plans for Level 3 features on MY 2021 A8 in April 2020, apparently due to regulatory uncertainty [Berman, 2020]). However, developmental deployments have been on the rise as ADS developers continue to test and improve their technology. In terms of *connected* vehicle deployments, a map of current (yellow) and planned (blue) deployments as of 2021 is illustrated in Figure 9.3, with a complete list included in Appendix A.

FIGURE 9.3 Current (yellow) and planned (blue) connected vehicle deployments throughout the United States as of 2021.

It should be noted that according to this map, *connected* is not necessarily linked to *automated*; however, the widespread deployment of connected vehicles demonstrates promise in the rapid growth of the associated technology. Like the expansion of connected vehicle deployments, the number of CAVs on public roadways has experienced substantial growth in recent years. A timeline of CAV current and projected deployments is depicted in Figure 9.4.

FIGURE 9.4 Timeline of current and projected CAV deployments in the United States as of 2021.

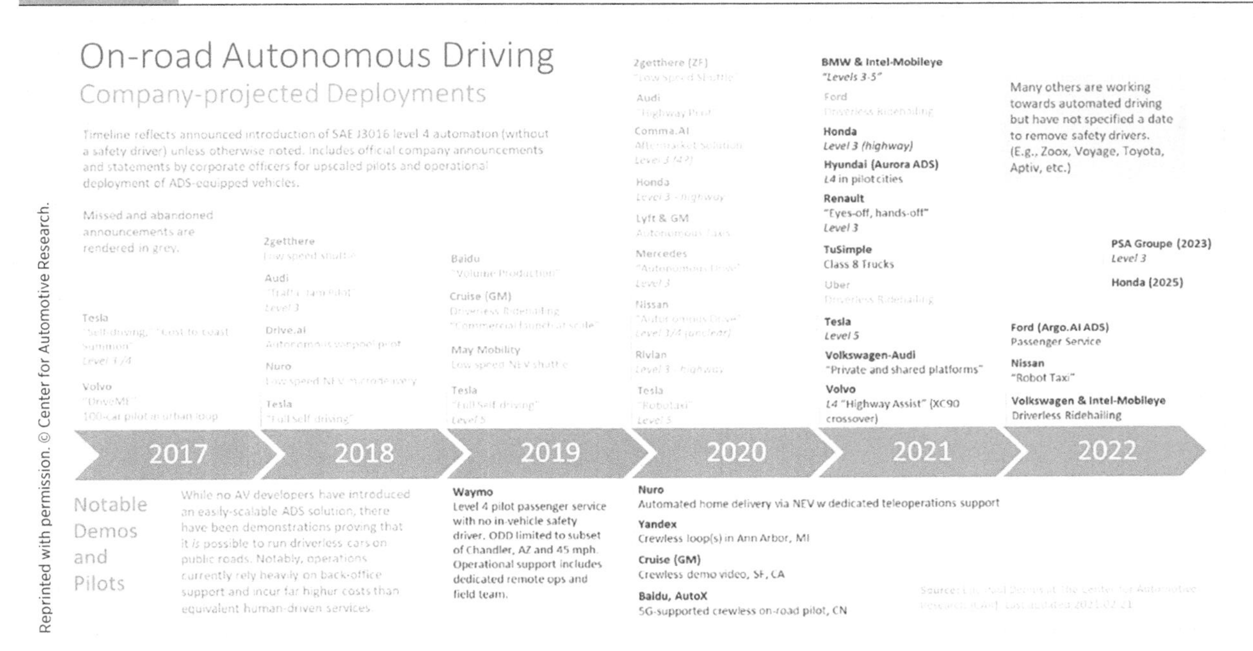

State of the Industry—Regulation and Legislation

> With more than 90% of serious crashes caused by driver error, it's vital that we remove unnecessary barriers to technology that could help save lives. We do not want regulations enacted long before the development of automated technologies to present an unintended and unnecessary barrier to innovation and improved vehicle safety."
>
> —(Owens, 2020)

There is a wide variance in OEM, government, consumer advocacy groups, etc. statements about Level 5 availability: will we approach Level 5 only asymptotically, i.e., keep getting closer but never actually achieve it? Even the implementation of Level 3 automation may be problematic since passing control of the DDT to a DDT Fallback-ready user is difficult to achieve in a sufficiently small time. Further yet, what is the appropriate timing for passing off control of the DDT if required to do so? While an NHTSA study suggests that passing control to a human DDT Fallback-ready user was not a problem (NHTSA, 2018), critics claim that the "quality" of the human driver fallback was not measured (DeKort, 2017). Even penetration of Level 2 automation is challenging due to the requirement that the driver must remain engaged in *monitoring* the DDT without actually *performing* any DDT task, potentially for extended periods of time. Existing research indicates that humans are generally poor monitors of automation (Pethokoukis, 2018), and there could be an increased risk of inattention and distraction. There remains much work needed to adapt traditional practices of vehicle regulation and evaluation to sufficiently assess the technology entering our transportation infrastructure.

In the United States, the U.S. DOT and NHTSA have made progress identifying the key challenges and goals associated with the introduction of CAVs into our transportation landscape in a series of publications. In 2017, the U.S. DOT released an initial groundwork for CAV advancement in "*Automated Driving Systems: A Vision for Safety 2.0*" (Published: September 12, 2017) in partnership with the NHTSA promoting best practices and highlighting the importance of proven safety in the development of automated transportation technology. Since then, the U.S. DOT has published three additional documents:

- "*Automated Vehicles 3.0: Preparing for the Future of Transportation*" Published: October 4, 2018
- "*Ensuring American Leadership in Automated Vehicle Technologies: Automated Vehicles 4.0*" Published: January 8, 2020
- "*Automated Vehicles Comprehensive Plan*" Published: January 11, 2021

These documents demonstrate an effort from the government to help guide the safe development of CAVs as they are deployed on public roadways, short of providing regulations and standards. In AV 3.0, the U.S. DOT addresses many of the challenges facing CAV deployment, benefits of the new technology, and plans for the government to assist in moving the industry forward through best practices and policy support, development of voluntary technical standards, targeted research, and modernization of regulations. While the temporary exemption granted to Nuro indicates some flexibility for the accommodation of CAVs within the current regulatory framework, it is only a preliminary step toward modernization of the regulations themselves. AV 4.0 is broken down into three primary goals each with several sub-categories including (National Science & Technology Council; United States Department of Transportation, 2020):

1. Protect Users and Communities
   a. Prioritize Safety
   b. Emphasize Security and Cybersecurity
   c. Ensure Privacy and Data Security
   d. Enhance Mobility and Accessibility
2. Promote Efficient Markets
   a. Remain Technology Neutral
   b. Protect American Innovation and Creativity
   c. Modernize Regulations

3. Facilitate Coordinated Efforts
   a. Promote Consistent Standards and Policies
   b. Ensure a Consistent Federal Approach
   c. Improve Transportation System-Level Effects

In the latest 2021 publication, *Automated Vehicles Comprehensive Plan*, the U.S. DOT explored a detailed path identifying the major questions surrounding the future of the transportation industry and addressed their plan to support the growth of these new technologies. An example of this support is illustrated in Figure 9.5, depicting zones involved in NHTSA's AV TEST initiative. The AV TEST initiative was launched in 2020 to increase transparency surrounding CAV testing across the transportation community and improve general societal awareness. This initiative is in line with promoting the support of public acceptance of CAVs on roadways, which is vital to the progression of the CAV industry and a prerequisite to widespread CAV deployment.

FIGURE 9.5 U.S. DOT CAV transparency and engagement for SAFE testing (AV TEST) initiative.

Reprinted from Public Domain.

The U.S. DOT's proposed path forward has been expressed throughout this series of publications and is summarized in Figure 9.6. The activities and initiatives relate to the three major goals set forth by the U.S. DOT:

1. Promoting collaboration and transparency between CAV developers, suppliers, and regulatory bodies.
2. Preparing the transportation system for CAV deployment.
3. Modernizing the regulatory environment to encourage CAV development.

These pillars are currently driving much of the activity supported by the U.S. government.

FIGURE 9.6 Flowchart summarizing the initiatives and activities of the U.S. DOT in support of the three overarching goals.

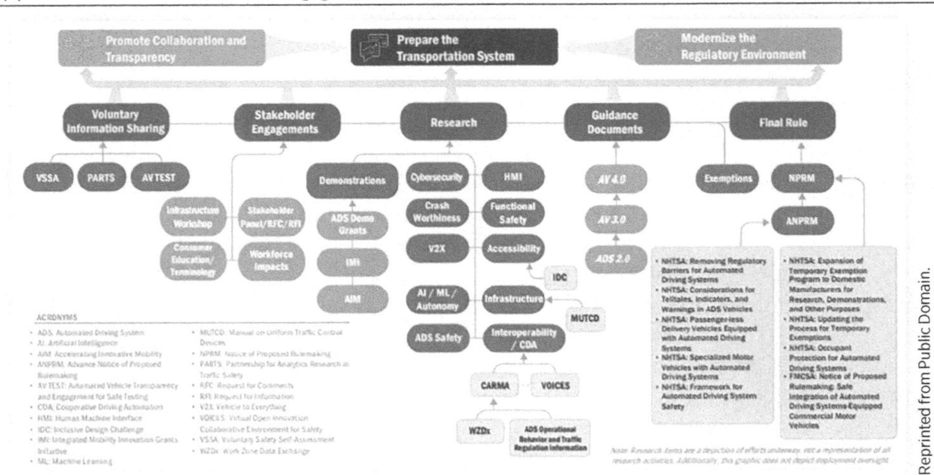

The most visible form of federal regulation is the VSSA introduced by the NHTSA. This assessment was introduced in the AV 2.0 document as a means of promoting transparency with OEMs testing vehicles on public roadways and as a venue for OEMs to present proof of concept for the safety of their vehicles without requiring the release of proprietary information. Although the VSSA is, as its name suggests, voluntary, many of the existing ADS developers have released a version of a VSSA in a variety of formats, with some being more detailed than others. A complete list of the developers who have submitted a VSSA to the NHTSA as of May 2021 is included in Table 9.1 (U.S. DOT, 2021a).

TABLE 9.1 List of OEMs with VSSA disclosures submitted to NHTSA.

Apple	Argo AI	Aurora	AutoX
BMW	EasyMile	Ford	GM
Ike	Kodiak	Local Motors	Lyft
Mercedes-Benz/Bosch L4-L5	Mercedes Benz L3	Motional	Navya
Nuro	Nvidia	Plus	Pony.ai
Robomart	Starsky Robotics	Toyota	TuSimple
Uber	Waymo	WeRide	Zoox

State Legislation and Regulation

As of 2020, there remained significant variation between U.S. states regarding rules and regulations surrounding CAVs. States have enacted legislation, executive orders, both, or neither. According to (NCSL, 2020), 16 states introduced legislation for CAVs in 2015, an increase from six states in 2012. By 2017, 33 states had introduced legislation. By 2018, 15 states had passed and enacted 18 CAV-related bills. Obviously, a patchwork of state legislation and regulation is not the ideal situation. Any CAVs being deployed would need to adhere to the particular local laws and regulations, which present much higher complexity and would require a significant amount of work to ensure compliance.

Standards Activities

Although the commercial availability and prevalence of CAVs on public roadways may be many years away, significant development has occurred in a short period of time on both the technology required to support such ambitious goals and the development of relevant standards and regulations for these vehicles. Various activities taking place within SAE International, ISO, UNECE, UL, and the Automated Vehicle Safety Consortium (AVSC) related to the research required to support the safe deployment and adoption of CAVs through appropriate standards and best practices. Some of these standards and best practices have been discussed in previous chapters, such as SAE J0316 and UL 4600.

The U.S. DOT recognizes that not all traditional regulations that once guided safety for the transportation industry are applicable when discussing CAVs. Certainly, it is not to say that CAVs should not meet similar crash safety standards and functional requirements; however, if a human driver will never be involved in the operation of a vehicle (i.e., the previously discussed example of Nuro), requirements for traditional driver-based features such as a side-view mirror or a steering wheel are fundamentally obsolete. For this reason, standards are being studied, research conducted, and new ideas proposed resulting from a variety of activities taking place across the world. As a step toward assessing ADS, entities such as ISO, NHTSA, European New Car Assessment Programme (EuroNCAP), and the IIHS have developed protocols and procedures to assess and rate the performance of Level 1 and Level 2 ADS. One of the early standards released by ISO relevant to automated systems is ISO 11270, titled "Intelligent transport systems - Lane Keeping assistance systems (LKAS) - Performance requirements and test procedures" (ISO, 2014). Although LKAS is defined as a driver assistance feature and not an automated feature (due to its intermittency), it is an early example of many standards to be released as they relate to the performance of automated functionality of vehicle safety. In order for these functions to truly improve the safety of vehicles on public roads, they must be implemented in a rigorously tested and proven manner, which is the purpose of standards committees.

Public Perception

Although implementation of the VSSA is certainly a step in the right direction when it comes to vetting vehicles prior to public deployment, there is still progress to be made. While many entities have released VSSAs, they remain voluntary, and the scope of many VSSA documents is limited, especially on a technical level. Most do not actually quantify how the vehicle achieves an acceptable level of safety or define what the level of safety achieved actually is. As the industry progresses, this proof will be vital to the public acceptance and trust in CAVs.

Numerous studies have been conducted to evaluate the public perception of CAVs and understand the relationship between CAV deployment and societal acceptance. One such effort, depicted in Figure 9.7, developed a parametric model to consider the public's acceptance of CAVs with several established control variables, including income, education, and car ownership (Yuen, Chua, Wang, Ma, & Li, 2020).

Results of this study indicated that these control variables had a surprisingly small effect on CAV acceptance. Factors with stronger correlation included attitude, behavioral

FIGURE 9.7 Parametric model employing theory of planned behavior to CAV acceptance.

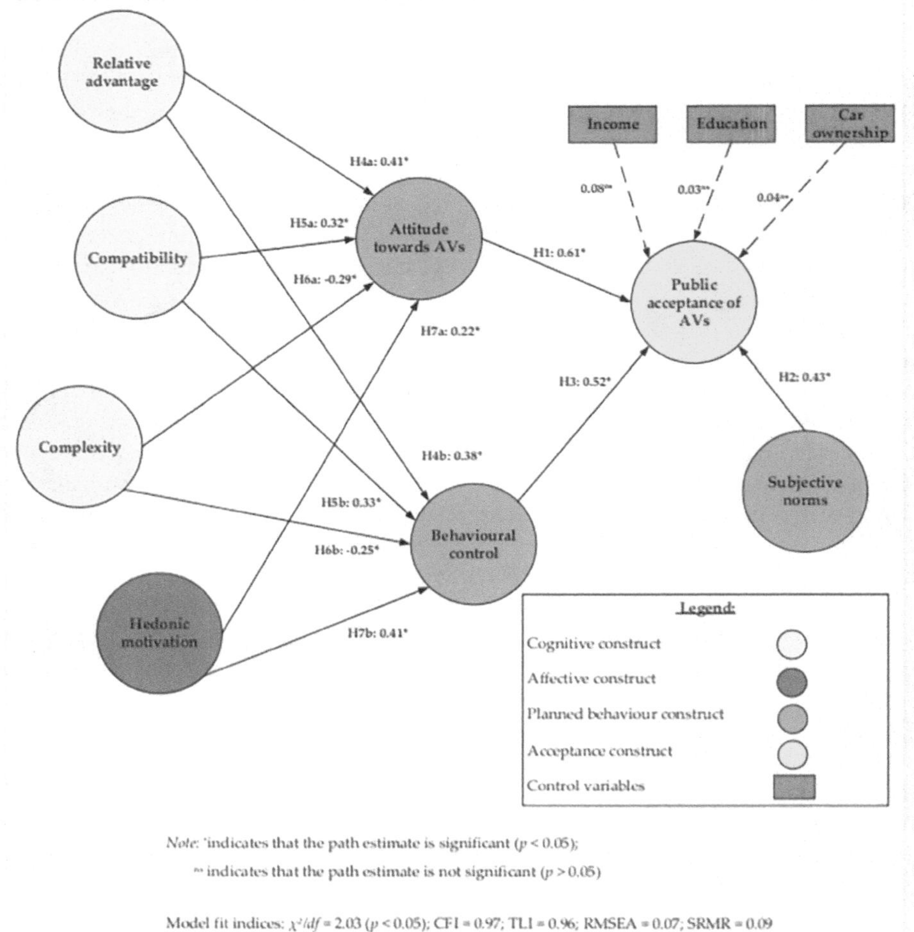

control, and subjective norms. Popular attitudes toward CAVs can be strongly impacted by the transparency of OEMs throughout the deployment process and positive/negative press about these deployments. Behavioral control relates to the barriers preventing the adoption of CAVs, including the high price tag associated with these vehicles, technological challenges that may limit the ODD or present unsafe scenarios, or the learning curve of using this new technology once it is deployed. Lastly, the subjective norms surrounding CAV deployment rely on media portrayal and social depiction. This implies that the negative media surrounding CAV incidents may serve to severely compromise public acceptance of this new technology. On the other hand, transparent V&V will strongly influence a more positive outlook on CAVs.

A poll conducted by the Partners for Automated Vehicle Education (PAVE) in early 2020 sought to understand the public perception of CAVs. This survey included 1,200 adults throughout the United States, approximately half of which owned cars with active safety features (PAVE, 2020). Figure 9.8 reveals societal attitudes toward CAVs, demonstrating a

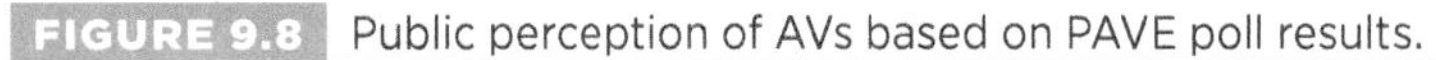

FIGURE 9.8 Public perception of AVs based on PAVE poll results.

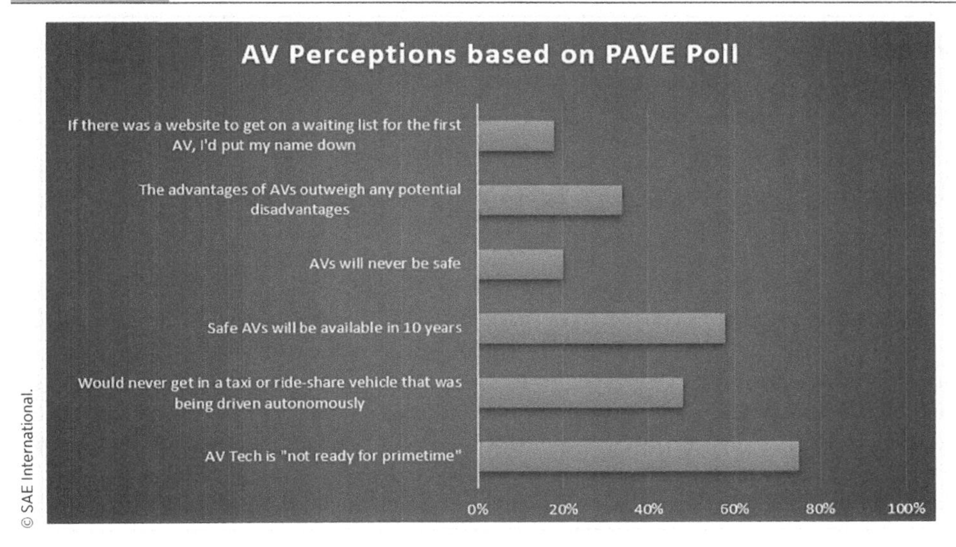

need to improve public trust in these vehicles, likely through testing and publication of test results.

Interestingly, the survey results indicated that most Americans know "nothing at all" or "a little" about the infamous Uber crash in Tempe, AZ (National Highway Traffic Safety Administration, 2018). However, the majority of respondents reported that a better understanding of the technology and even experiencing a ride in a CAV would improve their confidence in the technology (PAVE, 2020). Respondents who currently own vehicles equipped with ADAS features offered an overwhelmingly more positive outlook on CAV development, reinforcing results suggesting that a greater understanding and experience with the technology would facilitate public acceptance.

It follows from findings of studies on public attitudes toward CAVs that transparency on the validation and verification of their performance will be crucial for widespread acceptance. This is unlikely to be achieved through VSSA implementation alone and will likely require OEMs and regulatory bodies to share more comprehensive information and data on CAV performance, in a manner analogous to current NCAP standards established in the United States and around the world for today's consumer vehicles.

CAV-Related Research and Activities

As discussed in Chapter 8, V&V activities are underway within SAE including the development of information reports, recommended best practices, and standards. These documents are produced through discussions involving stakeholders throughout the industry ranging from OEMs to government agencies to third-party contributors. Through research being conducted within such committees and by participating members, the necessary dialogue is contributing to important conversations surrounding practical and safe methods for implementing a technology that will so greatly impact society, both CAV owners and traditional road users alike. These activities are being fueled through various mechanisms, including OEM contributions, university research, and government-funded initiatives. This section will explore some of the ongoing activities surrounding CAV research and developments, though it should be noted that there are far too many efforts to discuss them all here; thus, the authors have focused on a diverse set of topics currently being explored through university research collaborations with industry partners, some of which is being conducted by the authors themselves.

Advanced Vehicle Tracking through Infrastructure—IAM and ASU

As discussed in previous chapters, the Institute of Automated Mobility (IAM) was initiated as a result of an executive order in Arizona. The IAM has conducted extensive research in the area of CAVs through collaboration with governmental agencies in Arizona, Arizona universities including the University of Arizona (UofA), Arizona State University (ASU), and Northern Arizona University (NAU), in addition to industry partners such as Intel, Exponent, and State Farm. The initial phases of this research included developing a proposed set of operational safety metrics, which could be used to quantify the safety of CAVs (Wishart et al., 2020), evaluating these proposed metrics through simulation and collection of real-world data (Elli et al., 2021), and development of tracking algorithms capable of measuring these OSA metrics from existing infrastructure-based traffic cameras (Lu, et al., 2021).

As described in greater detail in Chapter 5, the ASU team has developed a vehicle tracking algorithm that is capable of collecting detailed vehicle trajectory information from existing infrastructure such as traffic cameras as shown in Figure 9.9. Analyzing the trajectory information obtained from the available data streams, the team has successfully applied the aforementioned operational safety metrics to obtain measures of vehicle performance. By applying such metrics to human-driven vehicles, comparisons can be drawn to existing traffic data for CAVs. Other companies are developing similar data processing pipelines for quantification of vehicle performance such as DERQ, Edge Case Research, and NoTraffic.

Deployment of Infrastructure-Based CAV Solutions—NAU and UofA

As previously discussed in Chapter 3, the U.S. DOT has contributed funding for research testbeds, one of which includes the SMARTDrive CV testbed located in Anthem, AZ. Data collection at this site has contributed to research involving the measurement of the aforementioned operational safety metrics proposed by the IAM. This research involved the analysis of infrastructure-based solutions which have been tested within this testbed including RADAR-based infrastructure solutions, traffic cameras, and LIDAR units, a sample of which is depicted in Figure 9.10. Data collection and analysis evaluated against ground-truth vehicle measurements acquired through differential GPS quantified the fidelity of various infrastructure-based solutions, which could be further evaluated based on cost, ease of implementation, and data management efficiency.

Infrastructure-based solutions are an important consideration for automation and especially connectivity as was demonstrated in detail through the V2X communication discussion in Chapter 3. Studies such as this one may facilitate the quantification of uncertainty, cost-benefit analyses, and data management to evaluate the optimal solutions for

FIGURE 9.9 Infrastructure-based vehicle tracking algorithm generating 3D-bounding boxes (top) and displaying vehicle trajectory information (bottom).

data collection and transfer. There are many solutions that may be applied to achieve sufficient automation and connectivity for CAVs, but it is important to consider solutions that may utilize existing infrastructure and technology as well as determine whether advances in these areas are necessary for an adequate solution.

Scenario-Based Testing of CAVs—Mcity and UMTRI

Mcity is a well-known test facility based at the University of Michigan in Ann Arbor and is regarded as one of the leading institutions in CAV research. In 2020, Mcity published a test method known as the ABC test for CAVs. This three-tiered approach focuses on **A**ccelerated testing, **B**ehavior competence, and **C**orner cases (Peng, 2020). These three facets are the major pillars currently guiding CAV testing due to the overwhelming time and cost needed to rigorously prove CAV safety. In a study by the RAND Corporation, it was determined that 8.8 billion miles would be required to prove CAV safety (Kalra & Paddock, 2016). One of the largest challenges facing the CAV community is tied to the impossibility of proving safety for a vehicle through such extensive testing. As such, Mcity has proposed the ABC test method to utilize scenario-based testing to ensure competency across a wide range of potential scenarios similar to a traditional driving test, accelerated testing to consider variations of commonly occurring scenarios in typical driving environments, and

FIGURE 9.10 Sample LIDAR data collected from Anthem testbed with corresponding traffic camera footage.

corner case evaluation to prove vehicle competency in rare, complex scenarios. Figure 9.11 demonstrates Mcity vehicles conducting test scenarios to be scored by the ABC test methodology.

FIGURE 9.11 Mcity testing of scenario-based vehicle competency.

This research is vital in establishing a feasible methodology to quantify safety for CAVs. However, the difficulty with scenario-based evaluations stems from the intelligence of CAVs as they are designed to learn through experience using AI and DNNs. This poses the issue that a vehicle may be "smart" enough to pass a defined driving test without actually understanding the fundamentals for the necessary driving behaviors. CAV safety evaluations are an imperative component needed to stabilize transparency and improve public trust during deployment.

CAV Interactions with Emergency Vehicles—VTTI

In 2020, the Virginia Tech Transportation Institute (VTTI) received a federal grant from the U.S. DOT providing funding for research designed to study the interaction between emergency vehicles and CAVs. In order to conduct such a study, VTTI is utilizing the

NVIDIA DRIVE Hyperion CAV test platform on public roadways in the Washington, D.C. area (Burke, 2020). The scope of this project includes studying various scenarios in which CAVs will be required to yield to emergency vehicles using connectivity technology as depicted in Figure 9.12.

FIGURE 9.12 Example depiction of CAV required to yield to oncoming emergency vehicle.

CAVs are required to follow all traffic laws applicable to traditional road users. Some of these laws are more complicated than others. For example, obeying the speed limit is simple in CAV programming, assuming the speed limit for any given area is known. This knowledge can be obtained through pre-existing HD map information or through signage recognition during driving. However, safely yielding to emergency vehicles can pose challenges depending on the surrounding traffic conditions. Deployment of CAVs creates uncertainty for emergency vehicles, which should be alleviated to minimize the impact on existing traffic infrastructure.

Significant research efforts are underway through a variety of institutions contributing to a better understanding of CAV technology and its impact on public roadways. The research discussed in the prior sections only begins to describe the multitudes of activities in progress. It should also be noted that CAV research and development is a global effort, though the focus of this chapter highlights activities within the United States.

What's Next?

There is a great deal of speculation and uncertainty regarding the future of CAVs within the United States and worldwide. As discussed within this chapter and throughout this book, significant progress has been made in CAV technology in recent years, with many companies and organizations around the world dedicated to the objective of CAV deployment. There is no reason to suspect that this progress will not continue. However, the industry consensus appears to be that, although the first 90% of the technological problem has been solved, and solved relatively quickly, that achieving an operational safety level that will be acceptable to all stakeholders is more difficult than originally imagined (Bloomberg, 2021). As research activities continue to inform CAV development, the industry will better understand the challenges and potential solutions available to optimize safety, drive down the cost of new technology, and facilitate the widespread deployment of CAVs on a global scale.

Standards and regulations surrounding the industry have lagged behind the technological advancements and will require greater focus to keep pace with the rapidly evolving industry. Governments and standards bodies appear to be starting to produce the necessary best practices, standards, and regulations. Coordination between all of the stakeholders will be important to avoid redundancy but also to leverage existing work so that the all-important regulations are implemented appropriately to provide for public safety while avoiding stifling industry innovation.

This book began with a discussion of the trends transforming the automotive industry and the observation that opinions on how the trends will change in the future vary widely, and where the automotive industry will actually end up is still unknown. The future of CAVs is certainly promising—the massive amounts of capital being invested, the stature of many of the companies and organizations involved, the significant attention devoted to the topic by government, and the growing interest of the general public all seem to assure that CAV development will continue. But the pace of deployments and commercialization, the order of use cases being deployed, and the overall impact on the transportation sector all remain to be seen. It is certainly an exciting time in the automotive industry, in no small part because CAVs present an opportunity for a fundamental shift in safety and mobility.

Appendix A

TABLE A.1 Operational connected vehicle deployments by state.

Operational Connected Vehicle Deployments by State	
Alabama	• University of Alabama, ACTION Advanced Transportation and Congestion Management Technologies Deployment (ATCMTD)
	• University of Alabama, Center for Advanced Vehicle Technologies and Alabama Department of Transportation (DOT)
Arizona	• Arizona Connected Vehicle Test Bed (Anthem)
California	• California Connected Vehicle Test Bed, Palo Alto
	• Prospect Silicon Valley Technology Demonstration Center Intelligent Transportation Systems (ITS) Lab
	• San Jose Connected Vehicle Pilot Study
Colorado	• Denver ATCMTD
	• U.S. RoadX Connected Vehicle Project
Delaware	• Delaware DOT Signal Phase and Timing (SPaT) Challenge Deployment

TABLE A.1 (Continued) Operational connected vehicle deployments by state.

Operational Connected Vehicle Deployments by State	
Florida	• Gainesville SPaT Deployment • Osceola County Connected Vehicle Signal Project • Pinellas County SPaT • Seminole County SR434 CV Deployment • Smart Work Zones • Tallahassee US 90 SPaT Challenge Deployment • Tampa Hillsborough Expressway Authority Connected Vehicle Deployment
Georgia	• City of Atlanta Smart Corridor Demonstration Project • Georgia DOT Connected Vehicle ATCMTD • Georgia DOT SPaT Project • Gwinnett County Connected Vehicle Project • I-85/"The Ray" Connected Vehicle Test Bed • Infrastructure Automotive Technology Laboratory • Marietta, GA, Emergency Vehicle Signal Preemption • North Fulton Community Improvement District
Hawaii	• Hawaii DOT Dedicated Short-Range Communications (DSRC) Deployment
Idaho	• Ada County Highway District
Indiana	• Indiana Connected Vehicle Corridor Deployment Project
Maryland	• I-895 Baltimore Harbor Tunnel DSRC • I-95 Fort McHenry Tunnel DSRC • US 1 Innovative Technology Corridor
Massachusetts	• Hope TEST
Michigan	• American Center for Mobility (Willow Run) • Ann Arbor Connected Vehicle Test Environment • Detroit ATCMTD • I-75 Connected Work Zone (Oakland County) • Lansing DSRC Deployment • Macomb County Department of Roads (MCDR) DSRC Deployment (MCDR/Sterling Heights Fire Department) • MCDR DSRC Deployment (Michigan DOT/General Motors SPaT Pilot) • MCDR DSRC Deployment (Michigan DOT/SMART Pilot) • Mcity Test Bed • Michigan DOT Wayne County Project • Michigan DOT I-94 Truck Parking Information and Management System • Road Commission for Oakland County DSRC • Safety Pilot Model Deployment • Smart Belt Coalition (MI) • Southeast Michigan Test Bed • U.S. Army Tank Automotive Research, Development, and Engineering Center "Planet M Initiative"

TABLE A.1 (Continued) Operational connected vehicle deployments by state.

Operational Connected Vehicle Deployments by State	
Minnesota	• Minnesota DOT DSRC • Roadway Safety Institute Connected Vehicle Test Bed
Nevada	• Las Vegas Freemont Street SPaT Corridor • I-580/Washoe County, NV
New Hampshire	• NHDOT SPaT, Dover
New Jersey	• City of New Brunswick Innovation Project • Integrated Connected Urban Corridor, Newark
New York	• New York City Connected Vehicle Project Deployment • New York State DOT Long Island Expressway INFORM I-495 Demonstration Test Bed • New York State Thruway Test Bed
North Carolina	• North Carolina DOT DSRC
Ohio	• City of Columbus—Smart City Challenge • NW US 33 Smart Mobility Corridor • Ohio Turnpike and Infrastructure Commission DSRC Project
Pennsylvania	• Pennsylvania Turnpike Harrisburg Connected Corridor • Pennsylvania DOT Harrisburg Demonstration • Pennsylvania DOT Ross Township Test Bed • Pennsylvania DOT SPaT Deployments and Test Beds • Philadelphia SPaT • Smart Belt Coalition • SmartPGH
Tennessee	• Tennessee DOT SPaT Challenge Project (Knoxville)
Utah	• Provo-Orem Bus Rapid Transit • Salt Lake Valley Snow Plow Preemption • Utah Transit Authority DSRC Traffic Signal Pilot Project
Virginia	• Fairfax County Connected Vehicle Test Bed • Virginia Smart Roads
Washington	• Washington State Transit Insurance Pool Active Safety-Collision Warning Pilot Project
Wisconsin	• Connected Park Street Corridor
Wyoming	• Wyoming Connected Vehicle Project Deployment

TABLE A.2 Planned connected vehicle deployments by state.

Planned Connected Vehicle Deployments by State	
Alabama	• University of Alabama ACTION ATCMTD
Alaska	• Alaska UTC
Arizona	• Loop 101 Mobility Project

TABLE A.2 (Continued) Planned connected vehicle deployments by state.

Planned Connected Vehicle Deployments by State	
California	• City of Fremont Safe and Smart Corridor • City of San Francisco Advanced Transportation and Congestion Management Technologies Deployment Initiative (ATCMTD) • Contra Costa ADS • Contra Costa ATCMTD • Freight Advanced Traveler Information System (FRATIS) • Los Angeles DOT Implementation of Advanced Technologies to Improve Safety and Mobility within the Promise Zone • San Diego 2020 ATCMTD
Colorado	• Colorado BUILD • Colorado TIGER • Colorado DOT Wolf Creek Pass ATCMTD • Denver ATCMTD Program
Delaware	• Delaware DOT ATCMTD
Florida	• ATCMTD I-4 Frame • Automated and Connected Vehicle Technologies for Miami's Perishable Freight Industry Pilot Demonstration Project • CAV Freight SR-710 • Central Florida AV Proving Ground • Connected Freight Priority System Deployment • Downtown Tampa AV Transit • I-75 Frame Ocala • Jacksonville BUILD • Lake Mary Boulevard CV Project • PedSafe Orlando • N-MISS • Pinellas County 2020 ATCMTD • SunTrax (Florida Turnpike) • University of FL Pedestrian and Bicycle Safety • US 1 Keys Coast • US 98 Smart Bay
Georgia	• CV-1K+ Project • GDOT 2020 ATCMTD • Gwinnett County CV Project
Hawaii	• Hawaii DOT C-V2X Project
Indiana	• Indiana DOT SPaT Deployment—Greenwood • Indiana DOT SPaT Deployment—Merrillville
Iowa	• Iowa City ADS
Kentucky	• Louisville TIGER
Maine	• Maine BUILD • Maine DOT 2020 ATCMTD

TABLE A.2 (Continued) Planned connected vehicle deployments by state.

Planned Connected Vehicle Deployments by State	
Michigan	• Detroit ATCMTD • MI ADS • MI BUILD • MI TIGER • Michigan DOT Intelligent Woodward Corridor Project • Smart Belt Coalition (MI) • University of Michigan 2020 ATCMTD
Missouri	• Kansas City US 69 Corridor SPaT Challenge • Springfield, MO SPaT Project • St. Louis SPaT Deployment Project
Nebraska	• NE ICM • NE TIGER
Nevada	• LV BUILD • RTC 2020 ATCMTD
New Jersey	• City of New Brunswick Innovation Project • Route (Rt) US 322 and US 40/322 Adaptive Traffic Signal (ATS) Project, Pleasantville, NJ • Rt 23, Rt 80 to CR 694 (Paterson Hamburg Turnpike), ATS C#1 • Rt 29, Rt 295 to Sullivan Way, ATS C#1, Hamilton Township and Trenton • Rt 38, Rt 70 to Union Mill Road, ATS C#1, Camden County • Rt 40, CR 606 to Atlantic Ave Intxn, Rt 50, Rt 40 to Cedar St ATS C#1, Atlantic City • Rt 46, Main St/Woodstone Rd (CR 644) to Rt 287, ITS, Parsippany-Troy Hills • Rt 46, Rt 23 (Pompton Ave) to Rt 20, ITS, Clinton Township • Rt 46, Rt 287 to Rt 23 • (Pompton Ave), ITS, Fieldsboro • Rt 73, Haddonfield Road to Delaware River, ATS C#2, Pennsauken Township Camden County • Rt. 1T and Rt. 440 by Communipaw Ave, Jersey City, ATS C#1, Jersey City • Rt. 18, Paulus Blvd to Rt. 287 SB Ramp, ATS C#2, Piscataway
New York	• Connected Region: Moving Technological Innovations Forward in the Niagara International Transportation Technology Coalition (NITTEC) Region
North Carolina	• NCDOT Multimodal CV Pilot
Ohio	• City of Columbus—Smart City Challenge • City of Columbus—Smart City Challenge • City of Columbus—Smart City Challenge • City of Columbus—Smart City Challenge • City of Columbus—Smart City Challenge • OH ADS • Smart Belt Coalition (OH)
Oregon	• OR ATCMTD

TABLE A.2 (Continued) Planned connected vehicle deployments by state.

Planned Connected Vehicle Deployments by State	
Pennsylvania	• PA ADS • Penn DOT I-76 Multi-modal Corridor Management Project • Penn Turnpike Harrisburg Connected Corridor
South Carolina	• South Carolina Connected Vehicle Test Bed
Tennessee	• Chattanooga Smart City Corridor Test Bed • Metro Nashville 2020 ATCMTD • TnDOT I-24 Corridor Nashville
Texas	• Arlington Cooper St. CV2X Project • Automated and Connected Vehicle Test Bed to Improve Transit, Bicycle and Pedestrian Safety • ConnectSmart—Houston • Dallas 2020 ATCMTD • Houston TIGER • Texas Connected Freight ATCMTD • TX ADS • TX I-10 ATCMTD
Utah	• UT DOT Connected Utah ATCMTD • Utah 2020 ATCMTD • Utah DOT CV Data Ecosystem Project
Virginia	• VA ADS • VA Port 2020 ATCMTD • VA Truck
Washington	• WSDOT SPaT Challenge (Poulsbo) • WSDOT SPaT Challenge Project (Spokane) • WSDOT SPaT Challenge Project (Vancouver) • WSDOT SPaT Projects in Lake Forest Park/Kenmore
Wyoming	• WY BUILD

References

Aalerud, A., Dybedal, J., & Subedi, D. (2020). *Reshaping Field of View and Resolution with Segmented Reflectors: Bridging the Gap between Rotating and Solid-State LiDARs*. Basel: Sensors.

Altekar, N., Como, S., Lu, D., Wishart, J., Bruyere, D., Saleem, F., & Head, L. K. (2021). Infrastructure-Based Sensor Data Capture Systems for Measurement of Operational Safety Assessment (OSA) Metrics. SAE World Congress.

Audi USA. (2016, August 15). Audi Traffic Light Information System. Retrieved from Youtube.com: https://youtu.be/OUxykbfmBEg

Autotalks.com. (n.d.). DSRC vs. C-V2X for Safety Applications.

Bekker, H. (2020, February 20). 2019 (Full Year) Europe: Best-Selling Car Models. Retrieved from Best-selling-cars.com: https://www.best-selling-cars.com/europe/2019-full-year-europe-best-selling-car-models/

Berman, B. (2020, May 4). Audi Gives up Plan for Hands-Off Autonomy for Next A8. Retrieved from Electrek.co: https://electrek.co/2020/05/04/audi-gives-up-plan-for-hands-off-autonomy-for-next-a8/

Bloomberg. (2021, August 22). Waymo Is 99% of the Way to Self-Driving Cars. The Last 1% Is the Hardest. Retrieved from Autoblog.com: https://www.autoblog.com/2021/08/22/waymo-is-99-of-the-way-to-self-driving-cars-the-last-1-is-the-hardest/

Burke, K. (2020, October 27). Listening to the Siren Call: Virginia Tech Works with NVIDIA to Test AV Interactions with Emergency Vehicles. Retrieved from NVIDIA: https://blogs.nvidia.com/blog/2020/10/27/virginia-tech-test-av-emergency-vehicles/

Business Wire. (2018, October 8). Siemens Mobility, Inc.'s Roadside Unit Is First to Receive OmniAir Certification. Retrieved from businesswire.com: https://www.businesswire.com/news/home/20181008005481/en/Siemens-Mobility-Inc.s-Roadside-Unit-is-First-to-Receive-OmniAir-Certification

Cadillac Customer Experience. (2017, March 9). V2V Safety Technology Now Standard on Cadillac CTS Sedans. Retrieved from Cadillac Pressroom: https://media.cadillac.com/media/us/en/cadillac/news.detail.html/content/Pages/news/us/en/2017/mar/0309-v2v.html

CB Insights. (2018, April 6). Where Auto Giants Are Placing Their Connected Car Bets. Retrieved from cbinsights.com: https://www.cbinsights.com/research/big-auto-connected-car-investments/

Cortright, J. (2017, July 6). Urban Myth Busting: Congestion, Idling, and Carbon Emissions. Retrieved from Streetsblog USA: https://usa.streetsblog.org/2017/07/06/urban-myth-busting-congestion-idling-and-carbon-emissions

DeKort, M. (2017, November 17). Autonomous Vehicles—We Are on the Wrong Road. Retrieved from LinkedIn: https://www.linkedin.com/pulse/autonomous-vehicles-we-wrong-road-michael-dekort/

Delphi Technologies. (n.d.). Making Sense of Sensors: Steering Angle Sensor. Retrieved from https://www.delphiautoparts.com/usa/en-US/resource-center/making-sense-sensors-steering-angle-sensor

Elli, M.S, Wishart, J., Como, S., Dhakshinamoorthy, S. et al. "Evaluation of Operational Safety Assessment (OSA) Metrics for Automated Vehicles in Simulation," SAE Technical Paper 2021-01-0868, 2021, doi:https://doi.org/10.4271/2021-01-0868.

Fiercewireless.com. (2020, November 18). FCC Votes to Open 5.9 GHz for Wi-Fi, C-V2X.

Gettman, D. (2020, June 3). DSRC and C-V2X: Similarities, Differences, and the Future of Connected Vehicles. Retrieved from Kimley-Horn.com: https://www.kimley-horn.com/dsrc-cv2x-comparison-future-connected-vehicles

Hawkins, A. J. (2020, March 4). Waymo's Next-Generation Self-Driving System Can 'See' a Stop Sign 500 meters Away. Retrieved from The Verge: https://www.theverge.com/2020/3/4/21165014/waymo-fifth-generation-self-driving-radar-camera-lidar-jaguar-ipace

Hill, C., & Krueger, G. (2012). ITS ePrimer—Module 13: Connected Vehicles. U.S. Department of Transportation - Intelligent Transportation Systems.

Howard, B. (2014, March 13). Audi A3 Is First Car with Embedded 4G LTE—But Will Owners Go Broke Streaming Movies? Retrieved December 4, 2020, from Extremetech: https://www.extremetech.com/extreme/178416-audi-a3-is-first-car-with-embedded-4g-lte-but-will-owners-go-broke-streaming-movies

ISO. (2014). ISO 11270:2014: Intelligent Transport Systems—Lane Keeping Assistance Systems (LKAS)—Performance Requirements and Test Procedures. ISO.

Jaillet, J. (2013, July 26). NTSB Asks for 'Connected Vehicles' Mandate. Retrieved from Commercial Carrier Journal: https://www.ccjdigital.com/business/article/14927741/ntsb-asks-for-connected-vehicles-mandate

Kalra, N., & Paddock, S. M. (2016). *Driving to Safety: How Many Miles of Driving Would It Take to Demonstrate Autonomous Vehicle Reliability.* RAND Corporation report (https://www.rand.org/pubs/research_reports/RR1478.html).

Korosec, K. (2021, May 5). GM CEO Mary Barra Wants to Sell Personal Autonomous Vehicles Using Cruise's Self-Driving Tech by 2030. Retrieved from techcrunch.com: https://techcrunch.com/2021/05/05/gm-ceo-mary-barra-wants-to-sell-personal-autonomous-vehicles-using-cruises-self-driving-tech-by-2030/

Lammert, M., Duran, A., Diez, J., Burton, K., & Nicholson, A. (2014). "Effect of Platooning on Fuel Consumption of Class 8 Vehicles Over a Range of Speeds, Following Distances, and Mass," *SAE Commercial Vehicle Congress.* Rosemont, IL, Paper 2014-01-2438.

Laws, J. (2014, November 1). Revving up V2V. Retrieved from Occupational Health & Safety Online: https://ohsonline.com/Articles/2014/11/01/Revving-Up-V2V.aspx

Li, Y. (2012). An Overview of DSRC/WAVE Technology. In Zhang, X., & Qiao, D., *Quality, Reliability, Security and Robustness in Heterogeneous Networks.* Berlin: Springer.

Litman, T. (2021). *Autonomous Vehicle Implementation Predictions: Implications for Transport Planning.* Victoria: Victoria Transport Policy Institute.

Lu, D., Jammula, V. C., Como, S., Wishart, J., Elli, M., Chen, Y., & Yang, Y. (2021). CAROM—Vehicle Localization and Traffic Scene Reconstruction from Monocular Cameras on Road Infrastructures. *International Conference on Robotics and Automation (ICRA).*

Maricopa County Department of Transportation. (2017). Connected Vehicle Test Bed: Summary.

Milanes, V., Shladover, S., Spring, J., Nowakowski, C., Kawazoe, H., & Nakamura, M. (2014). Cooperative Adaptive Cruise Control in Real Traffic. *IEEE Transactions on Intelligent Transportation Systems,* 15, 296-305. doi:10.1109/TITS.2013.2278494

Musk, E. (2019, April). On Using LIDAR. Tesla Launch Event.

National Highway Traffic Safety Administration. (2018). DOT HS 812 451—Quick Facts 2016.

National Science & Technology Council; United States Department of Transportation. (2020). *Ensuring American Leadership in Automated Vehicle Technologies: Automated Vehicles 4.0.* Washington, D.C.: United States Department of Transportation.

National Transportation Safety Board. (2019). Collision Between Vehicle Controlled by Developmental Automated Driving System and Pedestrian—Tempe, Arizona—March 18, 2018. NTSB/HAR-19/03—PB2019-101402.

NCSL. (2020, February 18). Autonomous Vehicles | Self-Driving Vehicles Enacted Legislation. Retrieved from NCSL: https://www.ncsl.org/research/transportation/autonomous-vehicles-self-driving-vehicles-enacted-legislation.aspx

NHTSA. (2015). Critical Reasons for Crashes Investigated in the National Motor Vehicle Crash Causation Survey. US Department of Transportation.

NHTSA. (2018). *Naturalistic Study of Level 2 Driving Automation Functions.* Washington, D.C.: NHTSA.

NHTSA. (2021a). AV TEST Initiative. Retrieved from NHTSA: https://www.nhtsa.gov/automated-vehicle-test-tracking-tool

NHTSA. (2021b, May). Voluntary Safety Self-Assessment. Retrieved from NHTSA: https://www.nhtsa.gov/automated-driving-systems/voluntary-safety-self-assessment

Owens, J. (2020, March 17). NHTSA Issued Proposals to Modernize Safety Standards. Washington, D.C.

PAVE. (2020, May). Pave Poll: Fact Sheet. Retrieved from Pave Campaign: https://pavecampaign.org/wp-content/uploads/2020/05/PAVE-Poll_Fact-Sheet.pdf

Peng, H. (2020). *Conducting the Mcity ABC Test: A Testing Method for Highly Automated Vehicles.* Ann Arbor: University of Michigan.

Pethokoukis, J. (2018, March 23). Why People Make Poor Monitors for Driverless Cars. Retrieved from AEI: https://www.aei.org/economics/why-people-make-poor-monitors-for-driverless-cars/

Qualcomm. (n.d.). C-V2X Products. Retrieved from qualcomm.com: https://www.qualcomm.com/products/automotive/c-v2x

SAE International. (2020, July). J2735_202007—V2X Communications Message Set Dictionary.

SAE International Surface Vehicle Recommended Practice, "Operational Safety Assessment (OSA) Metrics for Verification and Validation (V&V) of Automated Driving Systems (ADS)," SAE Standard J3237, In preparation.

SAE International Surface Vehicle Information Report, "Taxonomy and Definitions of Terms Related to Verification and Validation of ADS," SAE Standard J3208, In preparation.

Schmidt, B. (2021, April 27). No Need for Radar: Tesla's Big Jump forward in Self-Driving Technology. Retrieved from The Driven: https://thedriven.io/2021/04/27/no-need-for-radar-teslas-big-jump-forward-in-self-driving-technology/

Schrank, D., Eisele, B., & Lomax, T. (2019). Urban Mobility Report. Texas Transporation Institute.

Scribner, M. (2019, July 16). Authorizing Automated Vehicle Platooning, 2019 Edition. Retrieved from Competitive Enterprise Institute: https://cei.org/studies/authorizing-automated-vehicle-platooning-2019-edition/

Shepardson, D. (2019, April 26). Toyota Halts Plan to Install U.S. Connected Vehicle Tech By 2021. Retrieved from Reuters: https://www.reuters.com/article/autos-toyota-communication/toyota-halts-plan-to-install-u-s-connected-vehicle-tech-by-2021-idUSL1N22816B

Slovick, M. (2018, May 26). Toyota, Lexus Commit to DSRC V2X Starting in 2021. Retrieved from Innovation Destination - Automotive: https://innovation-destination.com/2018/05/16/toyota-lexus-commit-to-dsrc-v2x-starting-in-2021/

Taiwan Trade. (n.d.). V2X On-Board Unit with IEEE 1609.2/3/4 Stack Running on ThreadX RTOS, OBU-201U Enables Direct V2X Application Software Porting on the SDK. Retrieved from taiwantrade.com: https://www.taiwantrade.com/product/dsrc-v2x-on-board-unit-ieee-1609-x-protocol-stack-576209.html#

Tesla. (n.d.). Transitioning to Tesla Vision. Retrieved August 19, 2021, from Tesla.com: https://www.tesla.com/support/transitioning-tesla-vision

U.S. Department of Transportation. (n.d.). Vehicle-to-Infrastructure (V2I) Resources. Retrieved from Intelligent Transportation Systems - Joint Program Office: https://www.its.dot.gov/v2i/

U.S. DOT. (2021a). *Automated Vehicles Comprehensive Plan.* Washington, D.C.: U.S. DOT.

U.S. DOT. (2021b, March 31). Interactive Connected Vehicle Deployment Map. Retrieved from Transportation.gov: https://www.transportation.gov/research-and-technology/interactive-connected-vehicle-deployment-map

US Department of Transportation—National Highway Traffic Safety Administration. (2016). 49 CFR Part 571—Federal Motor Vehicle Safety Standards, V2V Communications—Notice of Proposed Rulemaking (NPRM).

US Department of Transportation. (2016). Smart City Challenge Lessons Learned.

Vardhan, H. (2017, August 22). HD Maps: New Age Maps Powering Autonomous Vehicles. Retrieved from Geospatial World: https://www.geospatialworld.net/article/hd-maps-autonomous-vehicles/

VTTI. (2021). An Examination of Emergency Response Scenarios for Automated Driving Systems. Retrieved from Virginia Tech Transportation Institute: https://featured.vtti.vt.edu/?p=1088

Walz, E. (2020, February 5). Audi Vehicles Can Now Communicate with Traffic Lights in Düsseldorf, Germany. Retrieved from FutureCar.com: https://www.futurecar.com/3766/Audi-Vehicles-Can-Now-Communicate-with-Traffic-Lights-in-Dsseldorf-Germany

White, A. (2020, December 15). The FCC Just Upended Decades of Research on Connected Vehicles. Retrieved from Car and Driver: https://www.caranddriver.com/news/a34963287/fcc-connected-cars-regulations-change-revealed/

Wishart, J., Como, S., Elli, M., Russo, B., Weast, J., Altekar, N., … Chen, Y. (2020). Driving Safety Performance Assessment Metrics for ADS-Equipped Vehicles. *SAE International Journal of Advanced and Current Practices in Mobility*, 2(5), 2881-2899. doi:https://doi.org/10.4271/2020-01-1206

World Health Organization. (2009). Global Status Report on Road Safety.

Yuen, K. F., Chua, G., Wang, X., Ma, F., & Li, K. X. (2020). Understanding Public Acceptance of Autonomous Vehicles Using the Theory of Planned Behaviour. *International Journal of Environmental Research and Public Health*, Volume 17, Issue 12: 4419.

Appendix B: Acronyms

Acronym	Definition
4WS	Four Wheel Steering
4G LTE	Fourth Generation Long Term Evolution
4WS	Four Wheel Steering
5G	Fifth Generation
AASHTO	American Association of State Highway and Transportation Officials
ACC	Adaptive Cruise Control
ACM	American Center for Mobility
ADS	Automated Driving System
AEB	Automated Emergency Braking
AI	Artificial Intelligence
AI	Artificial Intelligence
ASIL	Automotive Safety Integrity Level
ASU	Arizona State University
AUV	Autonomous Underwater Vehicle
AV TEST	Automated Vehicle Transparency and Engagement for Safe Testing
AVs	Automated Vehicles
AVSC	Automated Vehicle Safety Consortium
BIM	Basic Infrastructure Message
BRIEF	Binary Robust Independent Elementary Features
BSM	Basic Safety Message
BSW	Blind-Spot Warning
CAC	Criteria Air Contaminant
CACC	Cooperative Adaptive Cruise Control
CAROM	Cars on the Map
CAV	Connected and Automated Vehicle
CDA	Cooperative Driving Automation
CI	Crash Index
CIF	Criticality Index Function
CMU	Carnegie Mellon University
CNNs	Convolutional Neural Networks
ConOps	Concept of Operations
CPI	Crash Potential Index
CV	Connected Vehicle

Acronym	Definition
DARPA	(United States) Defense Advanced Research Projects Agency
DDT	Dynamic Driving Task
DFD	Data Feature Decision
DiL	Driver-in-the-Loop
DNNs	Deep Neural Networks
DOE	(United States) Department of Energy
DOT	(United States) Department of Transportation
DRAC	Deceleration Rate to Avoid a Crash
DSRC	Dedicated Short-Range Communication
DSS	Difference of Space Distance and Stopping Distance
ECU	Electronic Control Unit
EM	Electric Motor
EPS	Electric Power Steering
ESC	Electronic Stability Control
ESR	Electronically Scanning RADAR
EuroNCAP	European New Car Assessment Programme
EV	Electric Vehicle
EVSE	Electric Vehicle Supply Equipment
FAST	Features from Accelerated Segment Test
FCC	Federal Communications Commission
FCW	Forward Collision Warning
FMA	Fused Multiply-Accumulate
FMCW	Frequency Modulated Continuous Wave
FMVSS	Federal Motor Vehicle Safety Standards
FOV	Field of View
FSD	Full Self-Driving
GHG	Greenhouse Gas
GLOSA	Green Light Optimized Speed Advisory
GNSS	Global Navigation Satellite Systems
GPS	Global Positioning System
GPU	Graphics Processing Unit
HD Map	High-Definition Map
HiL	Hardware-in-the-Loop
HMI	Human-Machine Interface
HOV	High-Occupancy Vehicle
HSV	Hue-Saturation-Value
HUD	Heads-Up Display
IAM	Institute of Automated Mobility
ICE	Internal Combustion Engine
ICEV	Internal Combustion Engine Vehicle
IEEE	International Electrical and Electronics Engineers
IID	Independent and Identically Distributed
IIHS	Insurance Institute for Highway Safety
IM	Intersection Manager
IMA	Intersection Movement Assist
IMU	Inertial Measurement Unit

Acronym	Definition
IoU	Intersection over Union
ISO	International Organization for Standardization
ITS	Intelligent Transportation Systems
JDL	Joint Directors Laboratories
KLT	Kanade-Lucas-Tomasi (tracking algorithm)
LDA	Lane-Departure Warning
LIDAR	LIght Detection and Ranging
LKA	Lane-Keeping Assist
LTA	Left Turn Assist
MCDOT	Maricopa County Department of Transportation
MDP	Markov Decision Process
MEMS	Micro-Electromechanical system
MHT	Multiple Hypothesis Tracking
MiL	Model-in-the-Loop
MIT	Massachusetts Institute of Technology
ML	Machine Learning
MLP	Multilayer Perceptron
MPC	Model Predictive Control
MPrISM	Model Predictive Instantaneous Safety Metric
MRC	Minimal Risk Condition
MTC	Margin to Collision
MTT	Multi-Target Tracking
MTTC	Modified Time-to-Collision
MU	Measurement Uncertainty
N-FOT	Naturalistic-Field Operational Test
NaN	Not-a-Number
NHTSA	National Highway Traffic Safety Administration
NN	Neural Network
NOC	Network Operations Center
NTSB	National Transportation Safety Board
OBE	On-Board Equipment
ODD	Operational Design Domain
OEDR	Object and Event Detection and Response
OEM	Original Equipment Manufacturer
OOD	Out-of-Distribution
ORAD	On-Road Automated Driving (Committee)
ORB	Oriented FAST and Rotated BRIEF
OSA	Operational Safety Assessment
OTA	Over-The-Air
PATH	(California) Partners for Advanced Transportation
PAVE	Partners for Automated Vehicle Education
PET	Post-Encroachment Time
PICUD	Potential Index for Collision with Urgent Deceleration
PSD	Proportion of Stopping Distance
RADAR	RAdio Detection and Ranging
RCA	Radio Corporation of America

Acronym	Definition
RCNN	Region-based Convolutional Neural Network
ReLU	Rectified-Linear Unit
RGB	Red-Green-Blue
RGB-D	Red-Green-Blue-Depth
RSS	Responsibility-Sensitive Safety
RSU	Roadside Unit
RTK	Real-Time Kinematic
SCMS	Security Credential Management System
SD Map	Standard-Definition Map
SDO	Standards Development Organization
SFF	Safety Force Field
SiL	Software-in-the-Loop
SLAM	Simultaneous Localization and Mapping
SONAR	SOund Detection and Ranging
SOTIF	Safety of the Intended Functionality
SPaT	Signal Phase and Timing
TA	Time-to-Accident
TET	Time-Exposed Time-to-Collision
THW	Time Headway
TIDSS	Time-Integrated DSS
TIT	Time-Integrated Time-to-Collision
ToF	Time-of-Flight
TTC	Time-to-Collision
U of A	University of Arizona
UD	Unsafe Density
UL	Underwriters Laboratory
UMTRI	University of Michigan Transportation Research Institute
UNECE	United Nations Economic Commission for Europe
V&V	Verification & Validation
V2D	Vehicle-to-Device
V2G	Vehicle-to-Grid
V2I	Vehicle-to-Infrastructure
V2P	Vehicle-to-Pedestrian
V2V	Vehicle-to-Vehicle
V2X	Vehicle-to-Everything
VANET	Vehicular Ad Hoc Network
ViL	Vehicle-in-the-Loop
VRU	Vulnerable Road User
VSSA	Voluntary Safety Self-Assessment
WAVE	Wireless Access in Vehicular Environments
XiL	X-in-the-Loop

About the Authors

Jeffrey Wishart is Managing Engineer at the Test and Engineering Center of Exponent, Inc. as well as an adjunct professor in Automotive Systems of the Ira A. Fulton Schools of Engineering at Arizona State University. Dr. Wishart conducts research and development in the areas of energy and advanced transportation, including advanced powertrains, connected and automated vehicles, electric vehicle supply equipment, energy storage systems, and micromobility applications. Dr. Wishart is also Chair of the Verification and Validation Task Force under the On-Road Automated Driving SAE committee that is establishing standards for automated vehicles.

Dr. Wishart has over 20 years of experience in the advanced transportation and energy areas. In addition to academic and automotive industry positions, focusing primarily on testing and research of advanced powertrains, Dr. Wishart also worked for several years at a utility company in Queensland, Australia, conducting research into emerging energy technologies and also in asset management.

A Canadian, Dr. Wishart has a Ph.D. in Mechanical Engineering from the Institute for Integrated Energy Systems at the University of Victoria, an M.Sc. in Engineering Physics from the University of Saskatchewan, and a B.Sc. in Engineering Physics (Mechanical Engineering Minor) from the University of British Columbia.

Yan Chen received his B.S. and M.S. (with honors) in Control Science and Engineering from the Harbin Institute of Technology, China in 2004 and 2006, respectively. He received his second M.S. in Mechanical Engineering from Rice University in 2009, and a Ph.D. in Mechanical Engineering from Ohio State University in 2013. Dr. Chen is an Assistant Professor at Arizona State University. His research interests include design, modeling, control, and optimization of dynamic systems, specifically for connected and automated vehicles, electric vehicles, energy, and mechatronic systems. He is the author or co-author of more than 55 peer-reviewed publications. Dr. Chen serves as Associate Editor for IEEE Transactions on Vehicular Technology and IFAC Mechatronics, as well as IEEE CSS Conference Editorial Board. He is Vice Chair of the ASME Automotive and Transportation Systems Technical Committee. He is a recipient of the 2020 SAE Ralph R. Teetor Educational Award and 2019 DSCC Automotive and Transportation Systems Best Paper Award.

Steven Como is Senior Engineer at Exponent, Inc., where he specializes in accident reconstruction and the application of mechanical system and component analyses to vehicles. Mr. Como's experience investigating and analyzing motor-vehicle low- and high-speed frontal, rear, and side impacts includes conducting vehicle and site inspections, 3D scanning and photogrammetry techniques, computer simulation, and application of fundamental principles of physics. Mr. Como has received specialized training in accident reconstruction and has participated in physical demonstrations including low-speed crash testing, vehicle component testing, and crash testing of e-scooters. Additionally, Mr. Como is a certified Part 107 remote drone pilot and is certified as a professional engineer in the state of Arizona.

Mr. Como holds an M.S. and B.S. in Mechanical Engineering from Worcester Polytechnic Institute located in Worcester, Massachusetts, and is currently pursuing a Ph.D. in Systems Engineering from Arizona State University. Mr. Como's Ph.D. research is focused on automated vehicle technology, specifically in the development of methods to quantify the safety of automated vehicles.

Narayanan Kidambi is Senior Associate at Exponent, Inc., where he analyzes vehicle control systems, powertrain and propulsion systems, and vehicle emissions and energy use. His expertise lies in the intersection of mechanics, dynamics, and control. Dr. Kidambi holds an M.S. and Ph.D. in Mechanical Engineering from the University of Michigan and a B.S. in Mechanical Engineering from the University of California, Berkeley.

Dr. Kidambi has collaborated with industry partners to develop parameter estimation methods to improve the performance of driver assistance features and automated driving systems, to conduct research on uncertainty propagation in models of vehicle propulsion systems, and to develop adaptive structures for collaborative manufacturing robots and deployable systems. While pursuing his undergraduate degree, Dr. Kidambi was Subteam Lead for driver controls for the university's solar-powered vehicle racing team. He is broadly interested in transportation technologies and their wider societal impacts.

Duo Lu is an assistant professor of the Department of Computer Science at Rider University, where he specializes in automated vehicles, robotics, computer vision, and the application of motion capture to road transportation systems. Mr. Lu's expertise lies in the digitization of the motion of vehicles through visual perception sensors and reconstruction of traffic scenes for driving safety analysis and fine-grained traffic statistics. Mr. Lu holds a B.E. in Computer Science from Shanghai Jiao Tong University and M.S. in Computer Science from Arizona State University. Mr. Lu is currently pursuing a Ph.D. in Computer Science at Arizona State University.

Yezhou Yang is an assistant professor at School of Computing, Informatics, and Decision Systems Engineering, Arizona State University, directing the Active Perception Group. He has published over 60 papers in the fields of Computer Vision, Robotics, AI, NLP, and Multimedia, including papers at the top-tier conferences in these fields (CVPR, ECCV, ICRA, AAAI, ACL, and EMNLP) and journals in the related fields. His primary contributions lie in Computer Vision and Robot Vision, especially exploring visual primitives in interpreting peoples' actions and the scene's geometry from visual input, grounding them by natural language as well as high-level reasoning over the primitives for intelligent systems. He was a recipient of the NSF Career award in 2018, the Amazon AWS Machine Learning Research Award 2019, and the Qualcomm Innovation Fellowship in 2011.

Index

V

W

X